TILING GAMES

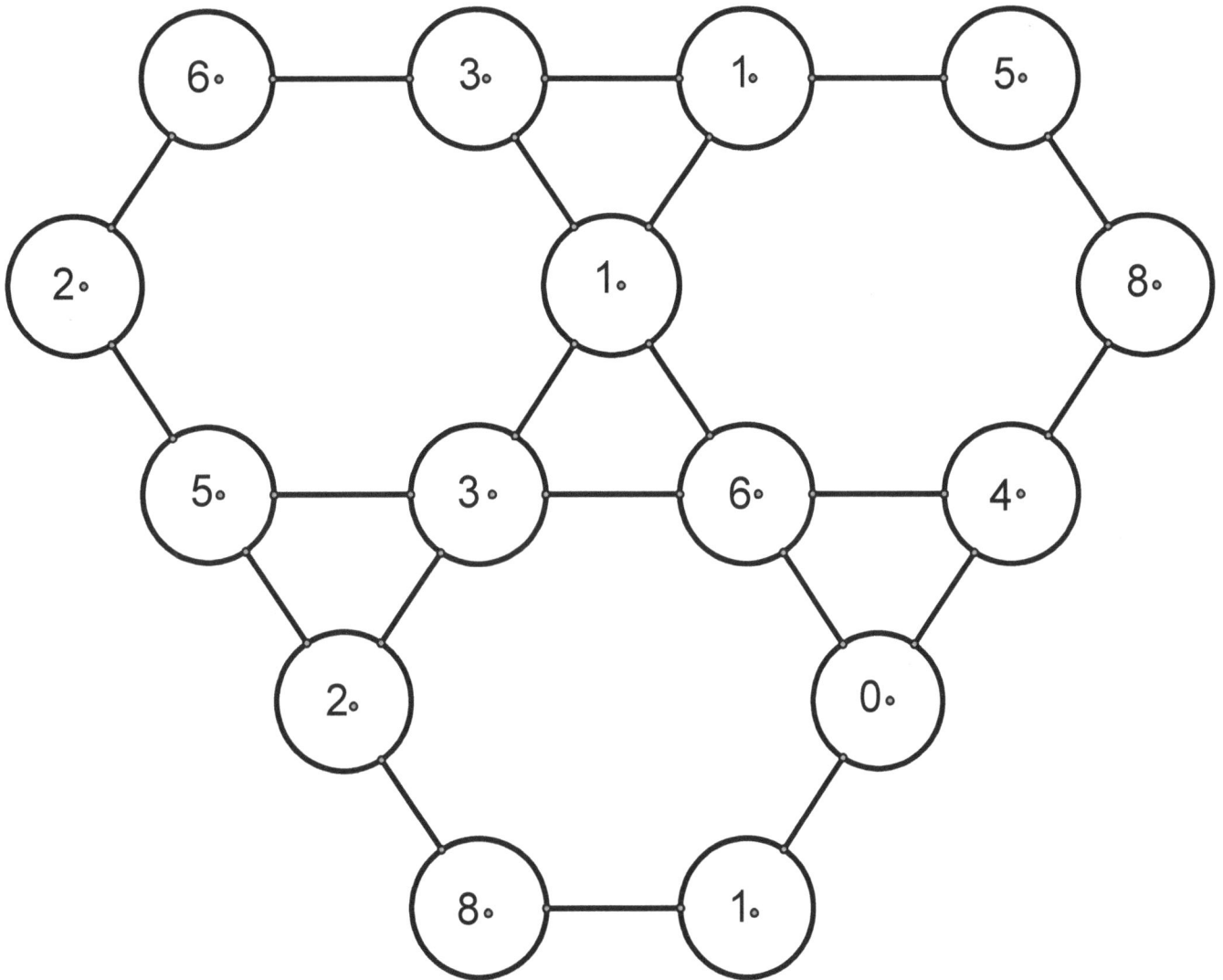

RAYMOND R. FLETCHER III

ISBN: 979-8-89465-016-6 (sc)
ISBN: 979-8-89465-017-3 (e)

Printed in the United States of America.

Integrity Publishing
39343 Harbor Hills Blvd Lady Lake,
FL 32159

www.integrity-publishing.com

CONTENTS

Preface. v

Chapter 1: Π_1-polygons 1

Chapter 2:   Nilpotent structures in the hexagon tiling 25

Chapter 3:   Nilpotent Π_3 subgraphs 63

Chapter 4: Π_4 polygons 71

Chapter 5:   The Π_5 Game 126

Chapter 6:   The Π_6 game 131

Chapter 7:   The Π_7 game 134

Chapter 8:   The game Π_8 137

Chapter 9:   Nilpotent polycubes 142

Chapter 10: Nilpotent Polygons in the Π_9 Plane 149

Chapter 11: Derivative planes 153

Chapter 12: The Π_{12} game. 162

Chapter 13: The Π_{13} game. 164

Chapter 14: The map of planes 169

Chapter 15: The Π_{15} game. 174

Chapter 16: The Π_{16} game. 177

Chapter 17: The Π_{17} game. 179

References 189

Index 191

PREFACE

The games described in this book are developed using the concept of *nilpotence*. To explain this concept we must first introduce the reader to the mathematical definition of a *graph*. The notation G = (V,E) is commonly used to represent a graph. The parameter V represents a collection of points called the *vertex set* of the graph G. The parameter E represents the *edge set* of G. The edge set is simply a collection of pairs of points in V; each such pair is called an *edge* of G. A graph can be drawn in the plane conveniently by representing each edge by a line or curve joining the two points which constitute the edge.

If the lines representing the edges of a graph can be drawn in the plane so that no two lines cross each other, then the graph is called *planar*, and in case the edges are so drawn the graph is called *plane*. In a plane graph the edges bound a collection of *regions*. Our games will take place on the vertex set of a plane graph. The plane graphs involved all arise from *tilings of the plane*. A tiling of the plane consists of a set of *tiles*, (in our games these tiles consist of a variety of polygons), which are placed edge to edge in such a way that the entire plane can, in principle, be covered.

Now suppose G = (V,E) is a plane graph, and Z represents the set of integers {…, -3, -2, -1, 0, 1, 2, 3, …}. Assign integers to the points in V so that the values assigned to the boundary points of each region of G sum to 0. We call this a *locally nilpotent assignment*. Every plane graph G has a locally nilpotent assignment since the integer 0 can just be assigned to each vertex. This is not an interesting assignment so we choose graphs for which a great variety of such assignments are possible. If for *every* locally nilpotent assignment the total value of all integers assigned to the vertices of a polygon P in G is 0, we say that P is *nilpotent*. The nilpotent polygons in our plane graphs will constitute *scoring configurations* in the associated game. In Figures (I, II) two nilpotent polygons in the tiling of the plane by squares are illustrated.

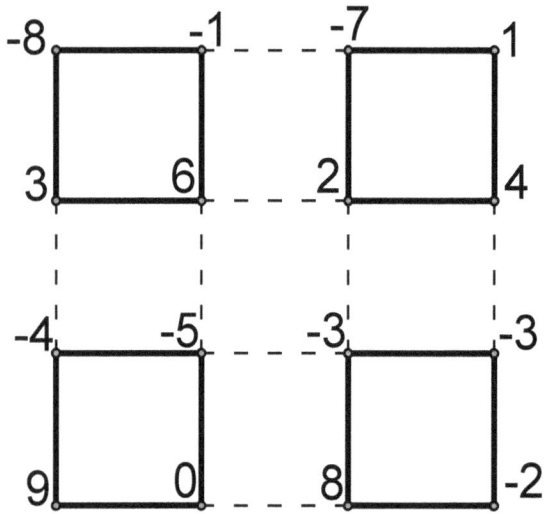

I

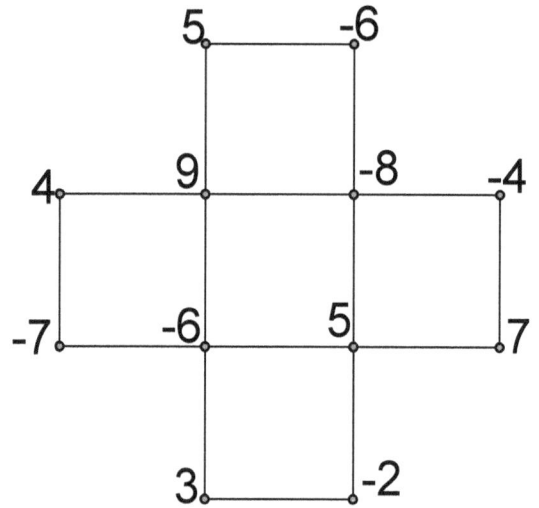

II

Both polygons have been provided a locally nilpotent assignment: the sum of values around each constituent square is 0. In Figure (I) the four squares indicated in bold constitute a *vertex cover* for the polygon P. That is to say that the squares are mutually disjoint and cover every vertex of P. The sum around each of these squares must be 0, and consequently the sum of all values in any locally nilpotent assignment for P must be 0. As a consequence P is nilpotent. It is easy to see that any rectangular polygon with an even number of points both vertically and horizontally is nilpotent.

The polygon in Figure (II) is also nilpotent, although it has no vertex cover by squares. The nilpotence can be understood by observing that edge values alternate in any linear sequence of squares. The tiling of the plane by squares is *monohedral*. This means that exactly one type of tile, the unit square, is used to tile the plane. The tiles belonging to any given tiling $\prod$ of the plane are *basic*. In any locally nilpotent assignment for $\prod$ the sum of integers in the boundary of each basic tile must be 0. In case a nilpotent polygon P in $\prod$ has a vertex cover consisting of basic tiles, we say that P is *composite*. In case P is nilpotent but has no such vertex cover, we say P is *prime*. Thus the polygon in (I) is composite, and the polygon in (II) is prime.

The determination of nilpotent polygons in a given plane Π is important for the construction of the game with base Π since these will correspond to scoring configurations. The determination of the prime or composite nature of nilpotent configurations is of theoretical interest. This interest is pursued especially in Chapters 1, 2, 3, 4.

The rules of play for all the games described in this book are the same except for some small variations. Discs with numbers in the range {0,...,9} are placed face down to constitute the *boneyard*. Each player draws 10 discs from the boneyard to constitute their *hand*. They are to be played on the vertices of a *board*. The board is derived from an appropriate section of a tiling Π of the plane by polygons. When a nilpotent polygon P of Π is completed with a total value which is a multiple of 5, the player earns a score equal to that value.

An integer is *congruent to 0 mod 5* in case it is a multiple of 5. For theoretical considerations nilpotence is defined over the set Z of integers, but for our games, nilpotence is defined by congruence to 0 mod 5. It is a basic fact of algebra that a sum of integer values which is 0 will also yield a sum equal to 0 mod 5. Thus our nilpotent polygons over Z will yield precisely polygons which count as scoring configurations in our games. In Figures (III, IV) the polygons in Figures (I, II) have been converted mod 5.

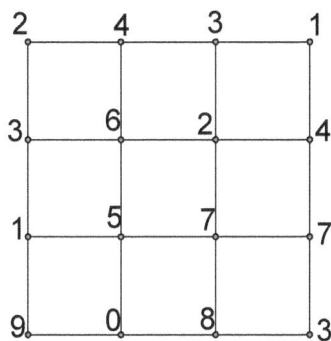

III

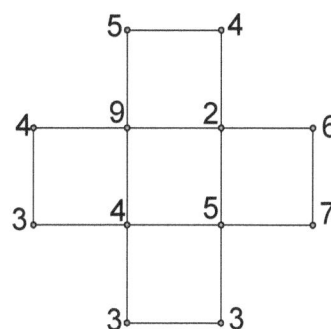

IV

The conversion is accomplished as follows. For example, if a is an integer in Figure (I), then choose a number b in the range {0, …, 9} such that a-b is *congruent to 0 mod 5*, i.e., a-b is a multiple of 5. Then replace a by b. There will always be two choices for b. For example -8 is congruent to both 2 and 7 mod 5. If a player were to complete the rectangle in Figure (III), he/she would score a total of 65 points, and if the cross in Figure (IV) were completed, a score of 55 points would be earned.

Most of the Theorems and some statements are left as exercises for the reader. Usually just a little basic algebra is required. There are even some research questions and unproved Conjectures which might prove of interest to the more mathematically inclined. The games themselves require little more than some mental arithmetic and geometric recognition. All the games have been designed so that they are fun to play.

The game boards can be constructed on matboard, and small wooden discs and number stickers can be found in a craft store. I drew a square lattice on the matboard with pencil, and then constructed the polygonal tiles using lattice points. An example of this technique is illustrated in Figure (V) for a polygon in the plane $\prod_9$.

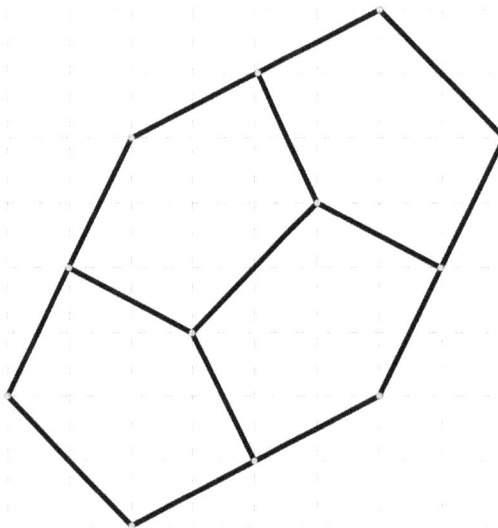

An index has been included which contains references for mathematical terms as well as special terms coined in the text. Special thanks go to my wife Sylvie for helping me to test the games.

Raymond R. Fletcher III

CHAPTER 1

Π_1-polygons

1.1 Introduction

In Mathematics a *graph G = (V,E)* is a set V of *vertices*, or *points* and a set E of pairs of points from V called *edges*. A graph can be represented by placing the points in a plane and connecting by a line or curve the pairs of points which form edges. A graph is *planar* in case it can be drawn with no crossing edges. In case it is drawn in this way, it is called a *plane graph*. An example of a plane graph is illustrated in Figure 1.1(I).

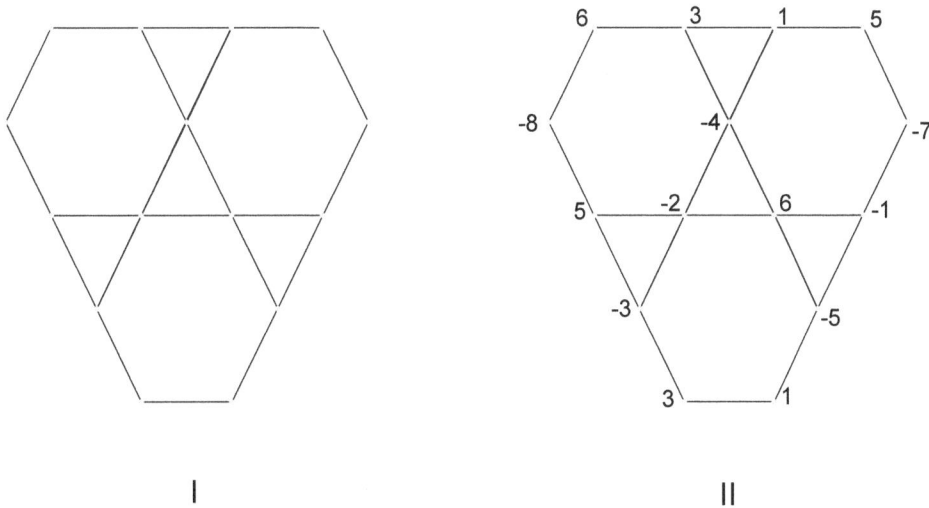

I   II

Figure 1.1

The graph in Figure 1.1(I) has 15 vertices, 21 edges and 7 *regions*. Regions in a plane graph are sections of the graph bound by edges and which contain no interior points. The regions for this graph consist of three hexagons and four triangles.

In Figure 1.1(II) the vertices of the same graph have been labeled with integers in such a way that the boundary points of each region have values which sum to 0. We call this a *locally nilpotent assignment from the set Z of integers*. Whenever a locally nilpotent assignment is made, (with the exception of the game Section(1.9)), it will be assumed that the assignment is made from Z. If for any locally nilpotent assignment, the sum of values assigned to the entire vertex set of a plane graph G is 0, we say that G is *nilpotent*. Our main objective for each of our games is to determine precisely which polygons in the associated tiling of the plane are nilpotent.

A section of a tiling, we call $\prod_1$, is illustrated in Figure 1.2.

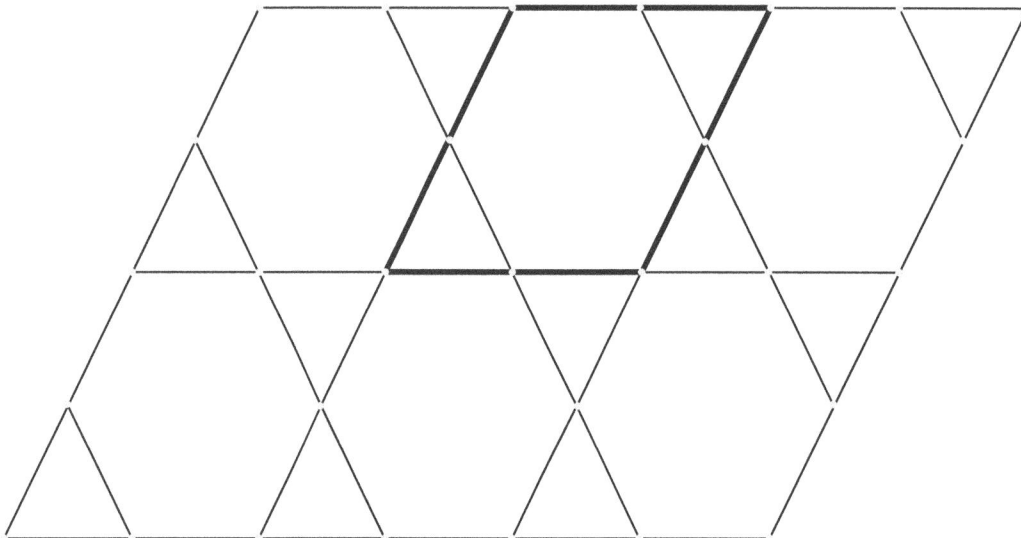

Figure 1.2

The tiling $\prod_1$ is *periodic* in the sense that it can be obtained by translations of one of its sections. The smallest such section is indicated in bold in Figure 1.2. The graph in Figure 1.1 is a polygonal *subgraph* of $\prod_1$. Polygonal subgraphs of $\prod_1$ will be referred to as *$\prod_1$ polygons*. We leave it as an exercise to show that this graph is nilpotent.

1.2 Triangles

In Figure 1.3 two $\prod_1$ triangles are illustrated. The notation Δ_m will be used for a $\prod_1$ triangle each side of which has m edges. The parameter m is always odd for triangles in the $\prod_1$ plane.

2

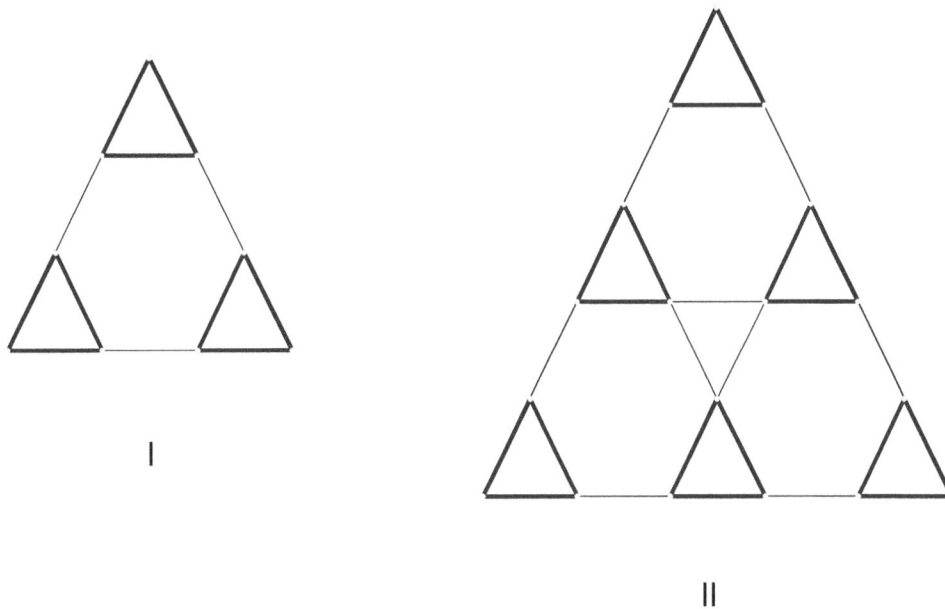

I

II

Figure 1.3

All $\prod_1$ triangles have a vertex cover consisting entirely of Δ_1 triangles, as indicated in Figure 1.3. As a consequence we have our first simple result.

Theorem 1.1: *Every $\prod_1$ triangle is nilpotent.*

1.2 Dual graphs

Given a plane graph G, the *dual of G,* denoted d(G) is a plane graph defined as follows. Place a point in the interior of each region of G, and join two such points in case they lie in regions of G which share a common boundary edge. Let these points and edges determine the vertex and edge sets of d(G). An example of this construction is given in Figure 1.4.

The underlying graph G in Figure 1.4 consists of rhombs placed so that every pair of adjacent rhombs form a chevron. G is a section of a tiling of the plane we refer to as the *Σ plane*. The section is a *Σ polyrhomb*. The dual of this polyrhomb, indicated in bold, is a $\prod_1$ polygon we call a *star*, and denote by J_1.

3

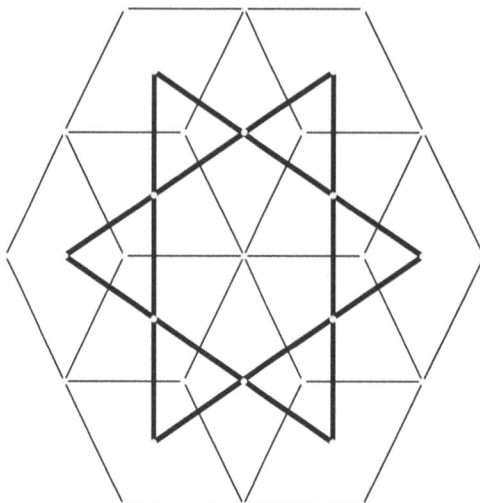

Figure 1.4

In fact the two planes are duals of each other: d(Σ) = $\prod_1$ and d($\prod_1$) = Σ. Properties of the Σ plane determined in [1] now carry over to the $\prod_1$ plane. We list these, leaving the proofs as exercises. A $\prod_1$ hexagon will be denoted by H(a,b,c,d,e,f) where the six parameters represent side lengths as encountered in a clockwise traversal of the hexagon boundary.

Theorem 1.2: *Let H(1,1,m,1,1,m), m odd, be a $\prod_1$ hexagon with a locally nilpotent assignment. Then the sum of the 6 corner values is 0.*

Let T(a,b,a+b) denote a $\prod_1$ trapezoid where a, a+b denote the lengths of the bases and b denotes the length of the lateral sides. In $\prod_1$ these are necessarily isosceles with the bases odd and the lateral sides even in length.

Theorem 1.3: *Let T(a,b,a+b) be a $\prod_1$ trapezoid with a locally nilpotent assignment. Then the sum of values of the corners of one base is equal to the sum of the corner values of the other base.*

Theorem 1.4: *If the $\prod_1$ triangle has a locally nilpotent assignment, then the sum of the three corner values is 0.*

Note that Theorem 1.1 and Theorem 1.4 together imply the following Corollary.

Corollary 1.5: *The $\prod_1$ hexagon H(1,m,1,m,1,m), m odd, is nilpotent.*

4

The $\prod_1$ plane is constructed entirely using three parallel classes of lines. If P is a $\prod_1$ polygon, then the intersection of a line with P is called a *diagonal of P*. A diagonal is *even* if it contains an even number of points of P, and *odd* in case it contains an odd number of points of P. A $\prod_1$ polygon is *even* in case each of its diagonals is even.

Theorem 1.6: *Let H be an even $\prod_1$ hexagon with a locally nilpotent assignment. Then the sum of the 6 corner values is 0.*

1.3 Circuits

A *path* in a graph is a sequence $(p_1 e_1 p_2 e_2 \ldots p_n e_n p_{n+1})$ of distinct points p_i and edges e_i with points p_i, p_{i+1} constituting the endpoints of edge e_i. In case $p_{n+1} = p_1$ a *circuit* is obtained. In a $\prod_1$ polygon a *segment* is a path which lies on a diagonal. A segment is *even* if it has an even number of points, and *odd* otherwise. In a $\prod_1$-polygon a circuit ρ is composed of segments. If each segment is odd then the circuit is *odd*, and if each segment is even, then the circuit is called *even*. The following results also derive from [1].

Theorem 1.7: *Let ρ be an odd circuit in a $\prod_1$ polygon with a locally nilpotent assignment. Then ρ has an even number of points. Moreover, If the corner values of ρ have values $(A_1, B_1, A_2, B_2, \ldots, A_n, B_n)$ in clockwise order, then $A_1 + A_2 + \ldots + A_n = B_1 + B_2 + \ldots + B_n$.*

The following Corollary is a consequence of Theorem 1.7 and the fact that the sides of any $\prod_1$ parallelogram have an odd number of points.

Corollary 1.8: *If P is a $\prod_1$ parallelogram with a locally nilpotent assignment, and (A_1, A_2), (B_1, B_2) are the two pairs of opposite corner points of P, then $A_1 + A_2 = B_1 + B_2$.*

Theorem 1.9: *Let P be a $\prod_1$ polygon with a locally nilpotent assignment, and let ρ be an even circuit in P. Then the number of corners of ρ is a multiple of 3, and the sum of the corner values is 0.*

In Figure 1.5, Theorems 1.7, 1.9 are illustrated in a $\prod_1$ hexagon.

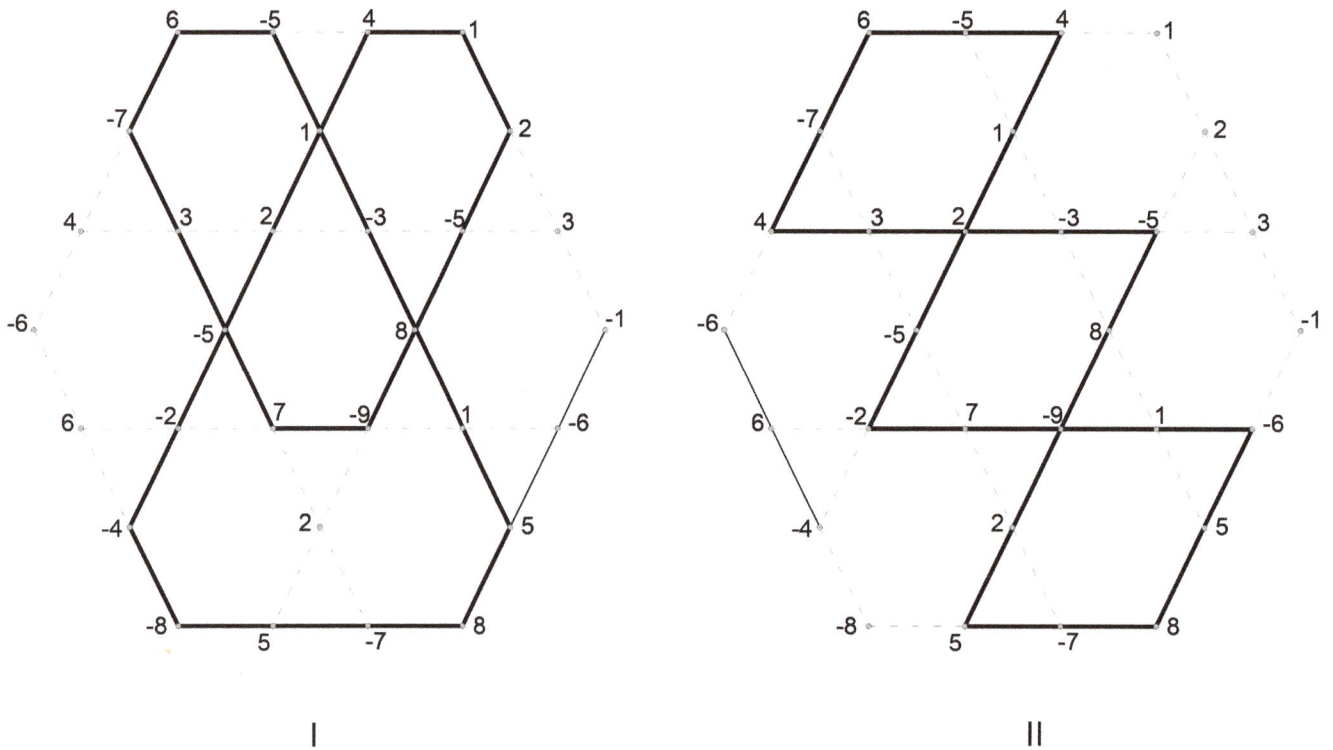

Figure 1.5

In Figure 1.5(I) an even circuit is illustrated with corner values {6,-5,5,8,-8,-4,4,1,2,-9,7,-7} which sum to 0 in accordance with Theorem 1.9. In Figure 1.5(II) an odd circuit in the same $\prod_1$ hexagon is indicated with successive corners {6,4,-2,-6,8,5,-5,4} whose alternate values each sum to 7 in accordance with Theorem 1.7.

1.4 Nilpotence

The following observation provides an easy way to determine if a given $\prod_1$ polygon is *nonnilpotent*.

Theorem 1.10: *If a $\prod_1$ polygon contains an odd diagonal then it is not nilpotent.*

The simple idea behind the proof of Theorem 1.10 is illustrated in Figure 1.6.

6

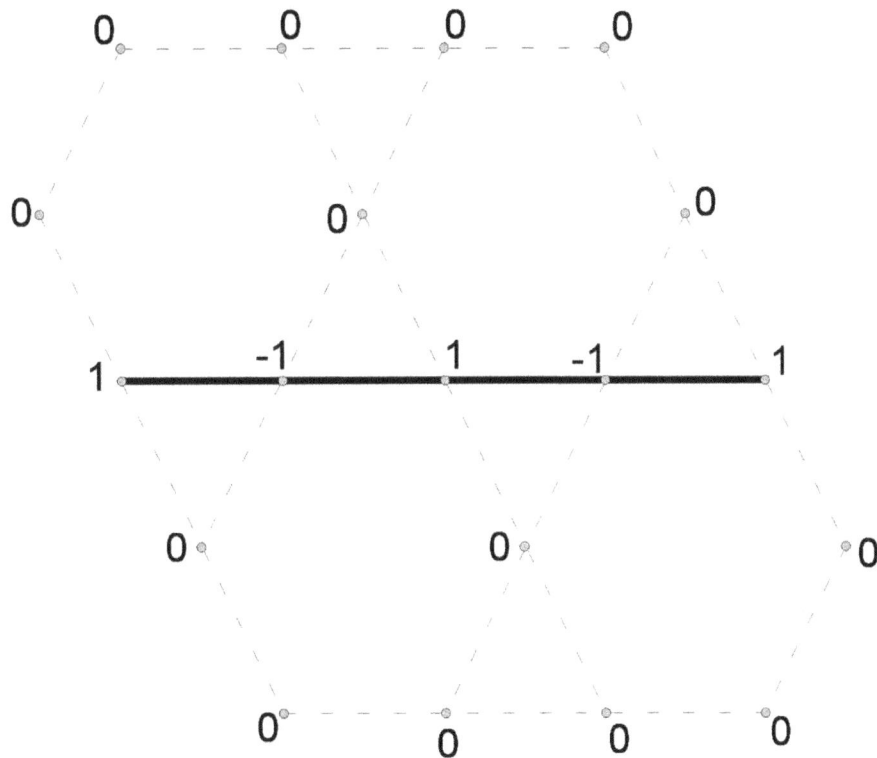

Figure 1.6

The H(1,3,3,1,3,3) Π_1 hexagon in Figure 1.6 has two odd diagonals, one of which is indicated in bold. To this diagonal we assign alternately the values 1,-1. The remaining points are assigned a value of 0. The result is a valid locally nilpotent assignment with total sum 1. It can thus be immediately concluded that this hexagon is nonnilpotent.

In case every diagonal, (including the sides), of a Π_1 polygon P is even, we say that P is *even*. Our main result concerning nilpotence is the following.

Theorem 1.11: *Every even Π_1 polygon is nilpotent.*

The idea behind the proof of Theorem 1.11 is illustrated in Figure 1.7.

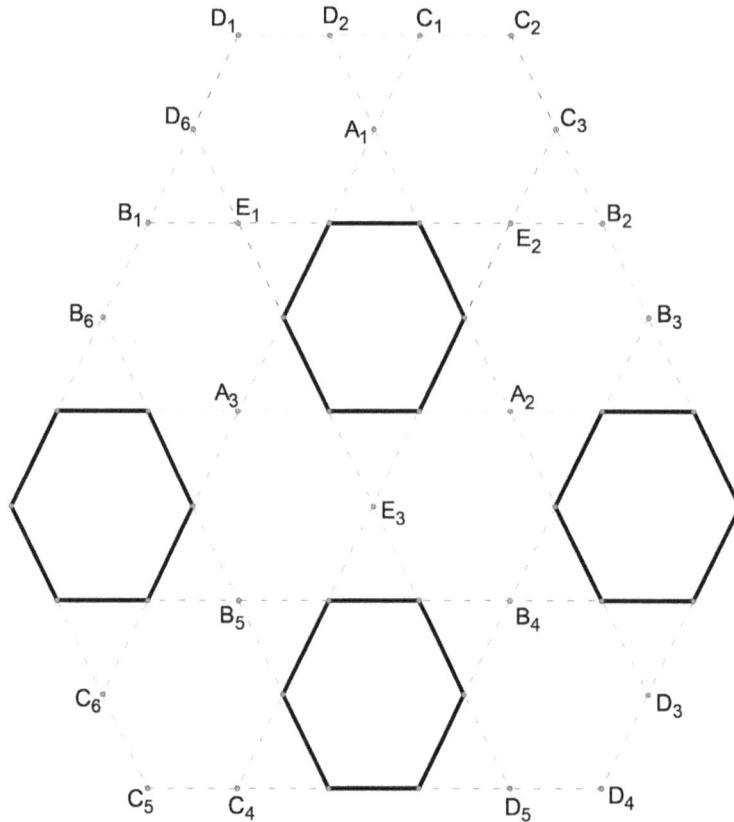

Figure 1.7

In Figure 1.7 we have a H(3,5,3,5,3,5) Π_1 hexagon whose vertex set has been partitioned into the corner point sets of even circuits. The partition consists of four H_1 hexagons in bold and corner points for even circuits A,B,C,D,E. Such a partition is always possible in case the base polygon is even. It then follows from Theorem 1.9 that this Π_1 hexagon is nilpotent.

In light of Theorems 1.10, 1.11 we can strengthen Theorem 1.11:

Theorem 1.12: *A Π_1 polygon is nilpotent iff it is even.*

A hexagon of the form P = H(a,b,a,b,a,b) is said to be *alternating*. In case P is a Π_1 hexagon, then a,b are both odd. Once it is observed that the only Π_1 hexagons which are even are the alternating ones, we obtain the following Corollary.

Corollary 1.13: *A Π_1 hexagon is nilpotent iff it is alternating.*

1.5 Prime and Composite

If a nilpotent $\prod_1$ polygon P has a dissection into smaller nilpotent polygons, we say P is *composite*, otherwise P is prime. Every $\prod_1$ triangle Δ_m with m≥3 is composite since it has a dissection into Δ_1 triangles. We will also say that Δ_m has a *{Δ_1}-factorization*. On the other hand, every nilpotent $\prod_1$ hexagon is prime. Composite $\prod_1$ polygons are easy to construct. Two examples are illustrated in Figure 1.8.

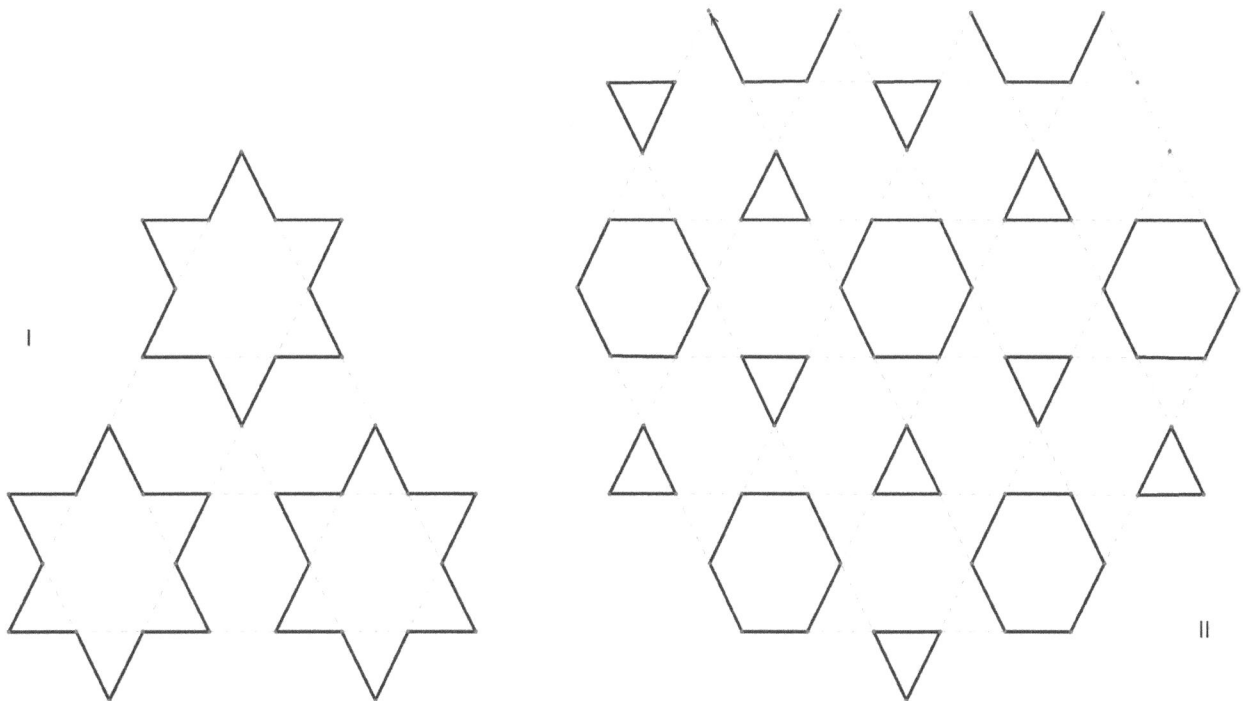

Figure 1.8

Both examples in Figure 1.8 can be extended to provide a factorization of the entire $\prod_1$ plane. The factors in Figure 1.8(I) are called *stars* and denoted by J_1. It is an easy exercise to show that J_1 is prime. We call the factorization a *{J_1}-factorization*. Inside this factorization numberless composite $\prod_1$ polygons can be found; we call them J_1 *chains*. The construction in Figure 1.8(II) extends to a *{H_1, Δ_1} factorization* of $\prod_1$. Unlimited numbers of composite $\prod_1$ polygons can be found here also. We shall refer to these as *Λ polygons*.

Stars larger than J_1 exist in the $\prod_1$ plane. These we denote by J_m, (m odd). The subscript m refers to the length of the line segments which form the boundary. In

Figure 1.9(I) the star J_3 is illustrated. In Figure 1.9(II) the *star cluster S_{12}* is illustrated. The subscript refers to the number of stars which can be found in the figure. In the interior of S_{12}, the smaller star clusters S_3, S_7 can be found. In both stars and star clusters all diagonals are even, so by Theorem 1.11, each is nilpotent. Both star constructions are also prime.

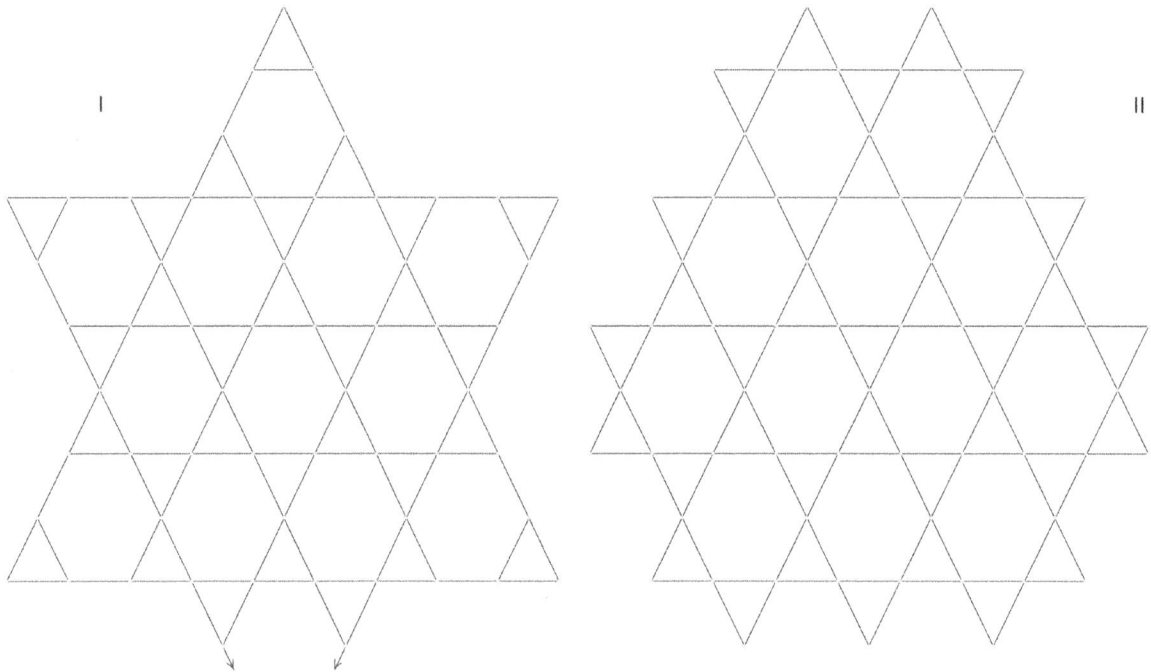

Figure 1.9

1.6 Circuit size

We have seen that a $\prod_1$ polygon P is nilpotent iff its vertex set has a partition into corner points of even circuits. For a given even circuit ρ, let the *size* of ρ, denoted $|\rho|$, be the number of corners in ρ. If θ is an even circuit partition in P, let $f(\theta)$ denote the maximum size of any circuit in θ, and let $\phi(P)$ denote the minimum value in the set { $f(\theta)$: θ is an even circuit partition in P}. The problem of interest is to determine $\phi(P)$ when P belongs to a particular class of $\prod_1$ polygons. Every triangle Δ_m has a Δ_1 factorization, so $\phi(\Delta_m) = 3$. For nilpotent $\prod_1$ hexagons we have the following result.

Theorem 1.14: *$\phi(H) = 6$ for any nilpotent $\prod_1$ hexagon H.*

In Figures 1.10, 1.11 constructions are illustrated that can be used to prove Theorem 1.14.

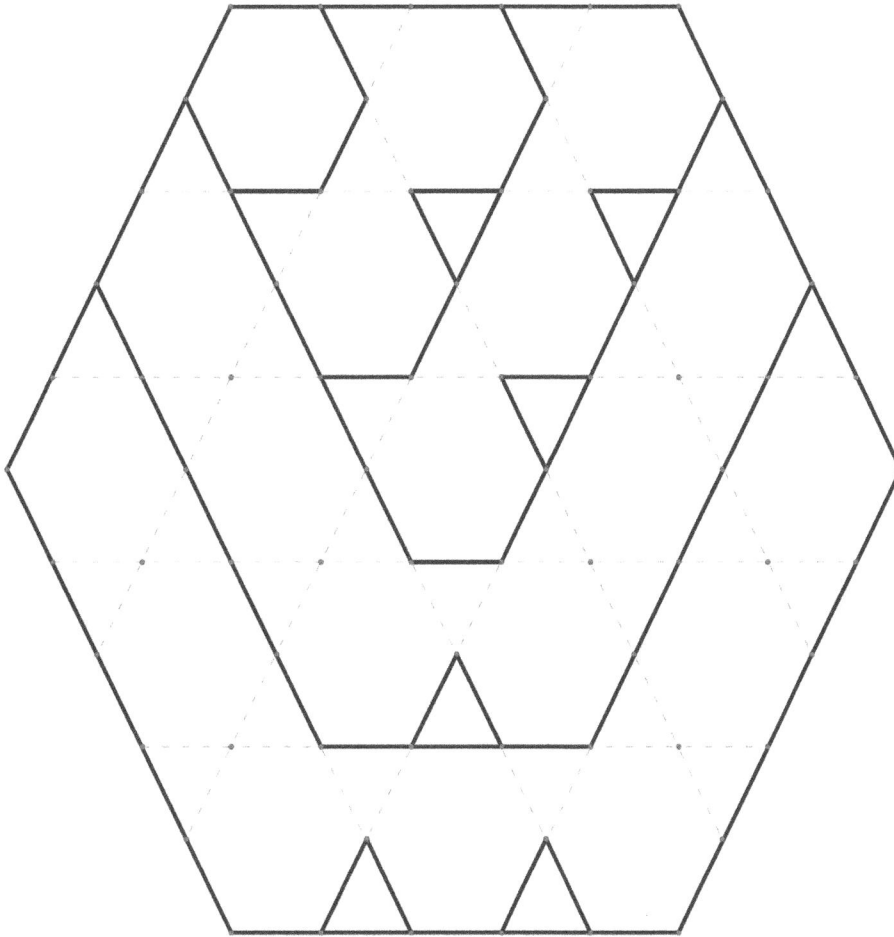

Figure 1.10

In both Figures the points not contained in the H_1 hexagon or any of the indicated Δ_1 triangles can be organized into corners of hexagonal even circuits. Both Figures illustrate the construction of *extensions*. Figure 1.10 illustrates the extension sequence: $H(1,3,1,3,1,3) \rightarrow H(1,5,1,5,1,5) \rightarrow H(3,5,3,5,3,5) \rightarrow H_5$, and Figure 1.11 illustrates the extension sequence: $H(1,3,1,3,1,3) \rightarrow H_3 \rightarrow H(3,5,3,5,3,5) \rightarrow H(3,7,3,7,3,7)$. The hexagonal circuits can be organized so that each member of the extension sequences are nilpotent. Also, an extension sequence can be constructed which yields any alternating $\prod_1$-hexagon H with $\phi(H) = 6$.

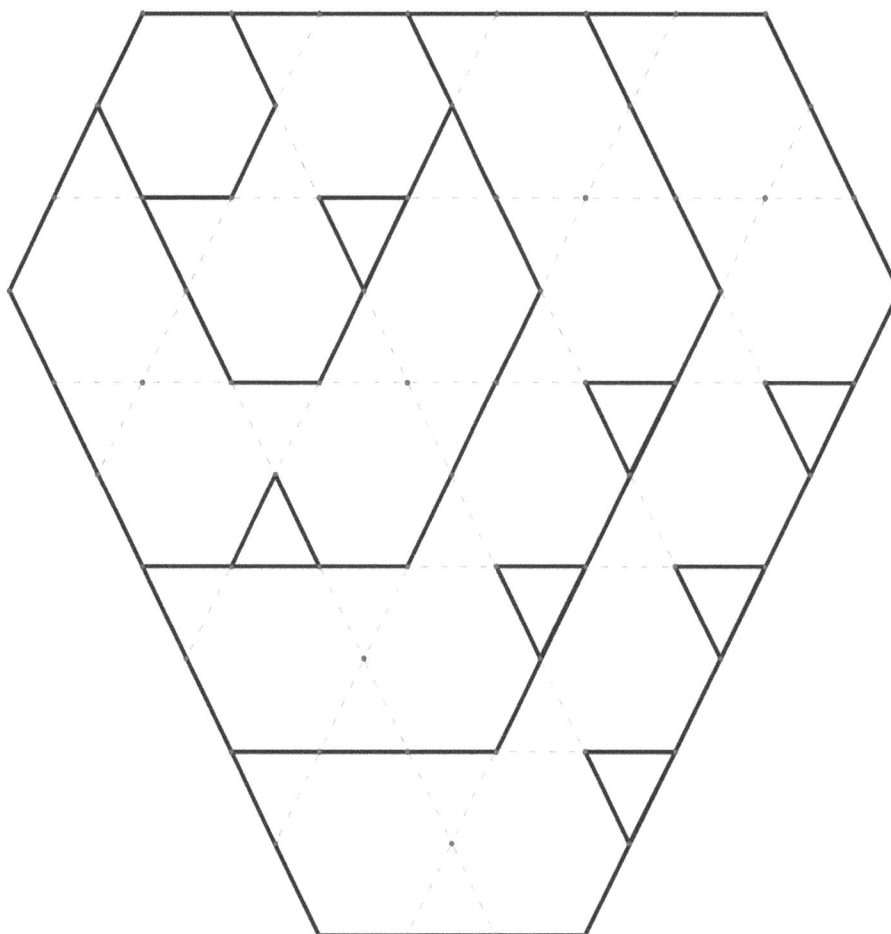

Figure 1.11

The proof of the following result we leave as an exercise.

Theorem 1.15: $\phi(J_m) = 6$ for $m \geq 5$, (m odd).

Determining even circuit structure for star clusters turns out to be quite challenging. We use the notation ΔS_m, m = 3, 6, 10, 15, …, to denote a star cluster whose constituent stars have a triangular formation. For these triangular star clusters it is easily seen that $\phi(\Delta S_m) = 3$. Consider, for example, the dissection of ΔS_6 into even 3-circuits in Figure 1.12.

General star clusters can assume very complicated formations and yet, it seems, that an even circuit dissection can always be found with each circuit having either 3 or 6 corners. In Figure 1.13 an attempt is made to tackle the problem by means of extensions.

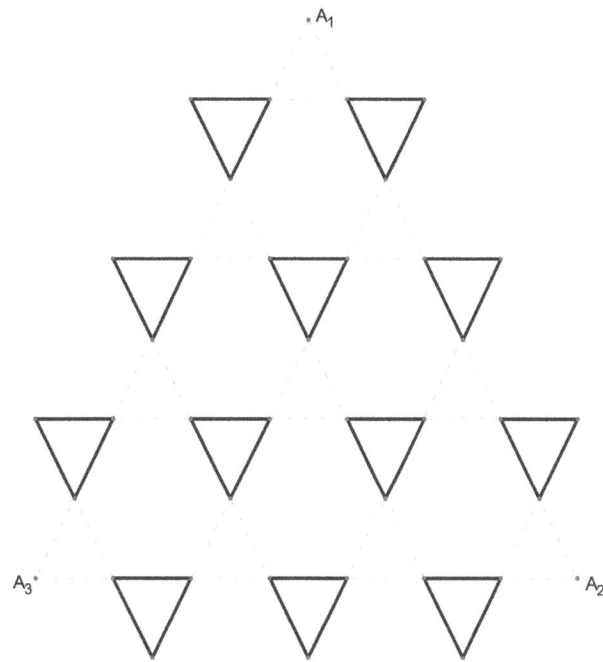

Figure 1.12

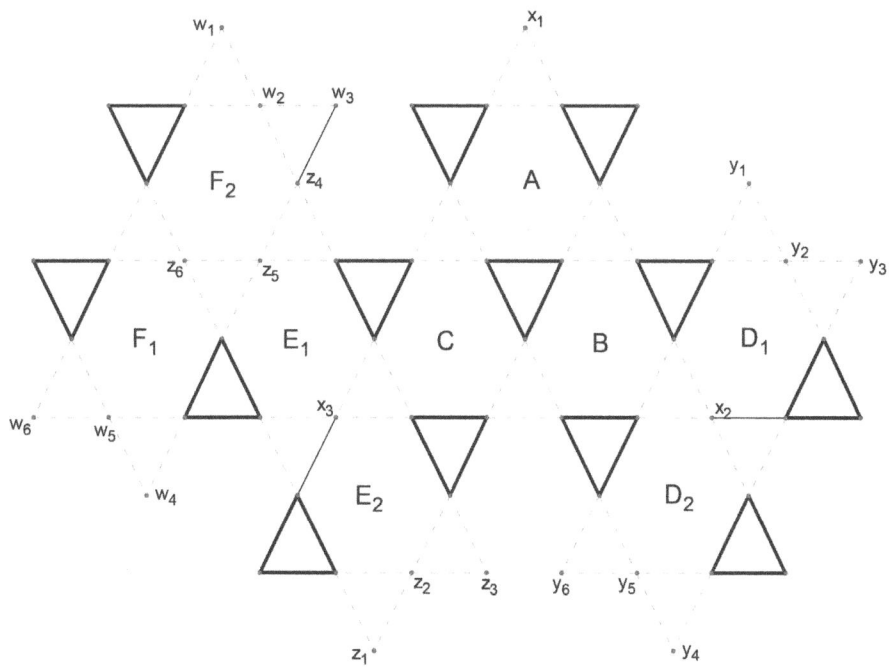

Figure 1.13

In Figure 1.13 the construction begins with an S_3 star cluster {A,B,C}. The extensions {D_1, D_2}, then {E_1, E_2}, and finally {F_1, F_2} occur in order. In the end result we easily obtain a circuit dissection in which each circuit has 3 or 6 corners.

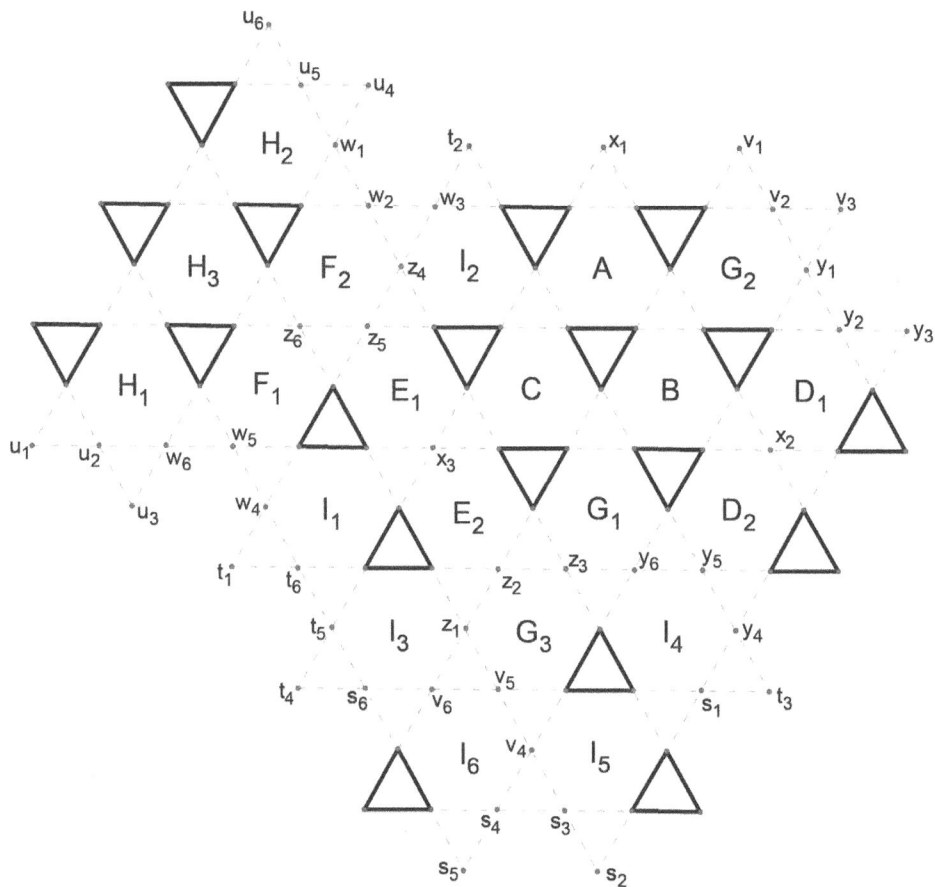

Figure 1.14

In Figure 1.14 the star cluster of Figure 1.13 is extended further with the G, H, and I extensions. We continue to obtain an even circuit dissection with circuits having 3 or 6 corners. The extensions can become quite complicated, for example the I extension beginning with star I_1 requires the addition of stars $I_2, \ldots, I_6$ in order to create a star cluster with all diagonals even. Due to this complexity we leave the following as an open problem.

Conjecture 1.16: If S_m is any nontriangular nilpotent star cluster, then $\phi(S_m) = 6$.

1.7 Soft $\prod_1$ polygons

A nilpotent $\prod_1$ polygon is *soft* if its boundary has no 60 degree angles. Every alternating or regular $\prod_1$ hexagon is soft. In Figure 1.15 an H(1,3,1,3,1,3) hexagon and an H(1,5,1,5,1,5) hexagon with several soft extensions are illustrated.

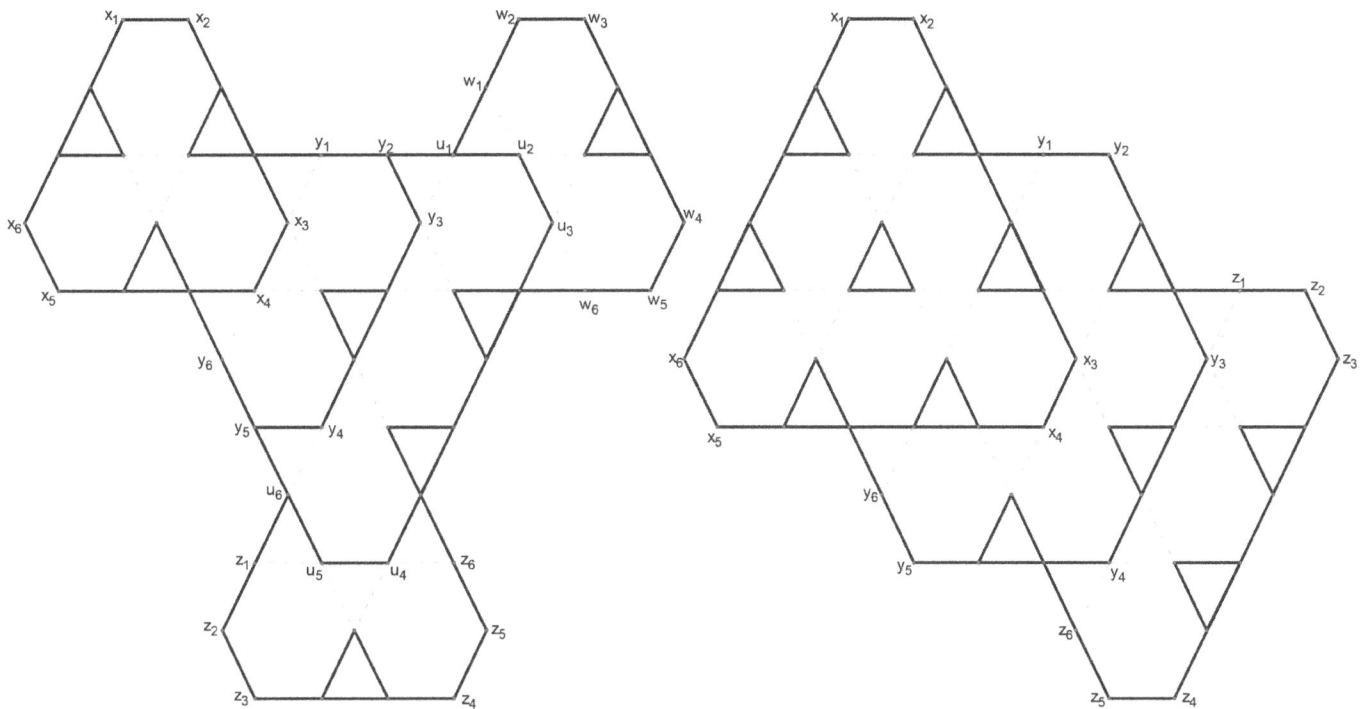

Figure 1.15

If a Δ_1 triangle is attached to the edges of each boundary hexagon of a soft $\prod_1$ polygon, a nilpotent star cluster is obtained, and conversely, if the boundary Δ_1 triangles of a nilpotent star cluster are removed, a soft $\prod_1$ polygon is obtained. Each soft polygon P in Figure 1.15 yields $\phi(P) = 6$, and as yet no nilpotent $\prod_1$ polygons have been constructed with $\phi(P) > 6$. We therefore have the following open question.

Question 1.17: *Does there exist a nilpotent $\prod_1$ polygon P with $\phi(P) > 6$?*

1.8 Conditional nilpotence

The $\prod_1$ polygons in Figure 1.16 each contain odd diagonals which are dashed. By Theorem 10 none of these polygons are nilpotent. In Figures 1.16(I,II,III)

15

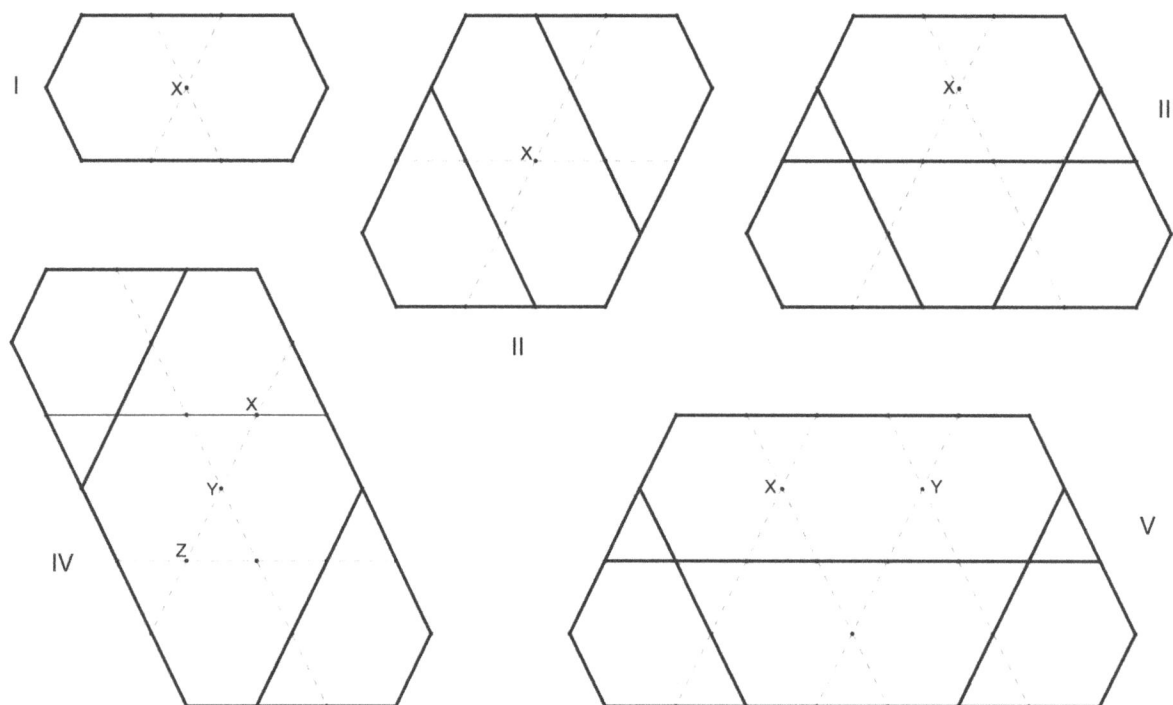

Figure 1.16

a single point labeled x marks the junction of the only pair of odd diagonals. In case x = 0, we leave it as an exercise to show that these hexagons become *conditionally nilpotent*. That is to say that in any locally nilpotent assignment with the requirement x = 0, the sum of values assigned to the full vertex set of the hexagon is 0. What we have hitherto called nilpotent will be sometimes called *intrinsically nilpotent* in order to distinguish this concept from conditional nilpotence.

In Figure 1.16(IV) the condition for nilpotence is x+y+z = 0, and in Figure 1.16(V) the condition for nilpotence is x+y = 0. The reader should check these assertions. Note that in Figure 1.16(IV) three points {x,y,z} are required to cover the four odd diagonals, and in Figure 1.16(V), two points {x,y} are required. The sets {x,y,z}, {x,y} and {x} in these examples are called *0-sets* since the sum of the member(s) must be 0 to insure conditional nilpotence. It is important that every odd diagonal be covered by an odd number of points in the 0-set, and that an even number of points in the 0-set cover an even diagonal. We are interested in 0-sets with minimum cardinality. Two more examples of conditionally nilpotent Π_1 polygons are illustrated in Figure 1.17.

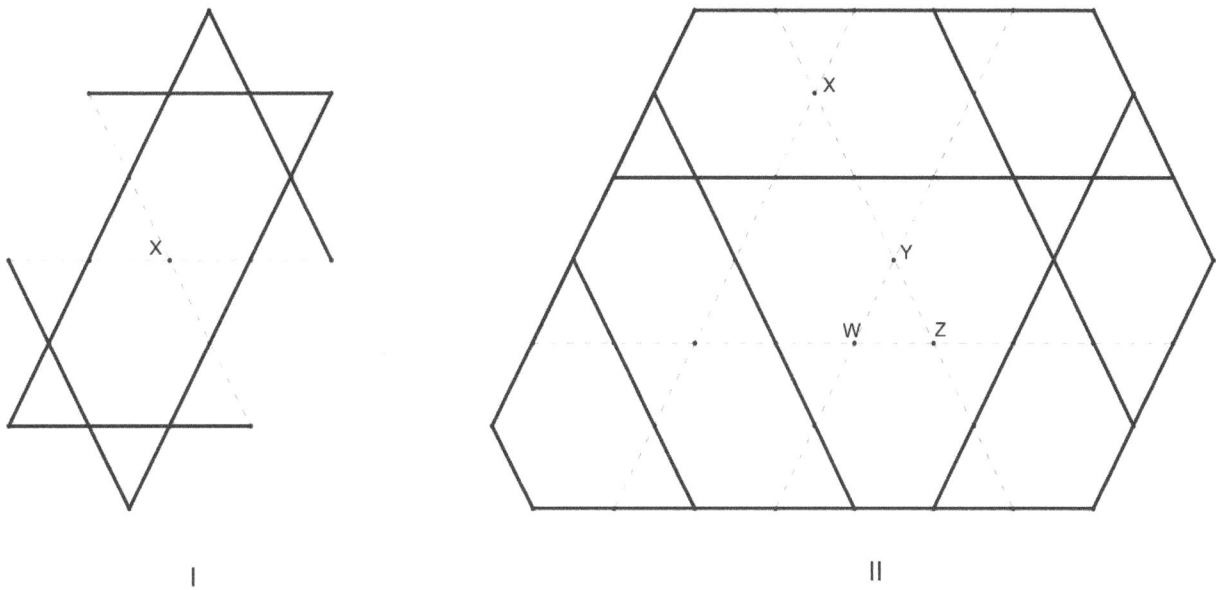

Figure 1.17

The construction in Figure 1.17(I) is the star cluster S_2. It is nilpotent iff $x = 0$. In Figure 1.17(II) the $\prod_1$ hexagon $H(1,5,5,3,3,7)$ is illustrated. It is nilpotent iff $x+y+z = 0$. Note that while the points $\{x,w\}$ cover all the odd diagonals, the relation $x+w = 0$ does not insure nilpotence.

1.9 The game $\prod_1$

A section of the plane $\prod_1$ provides the playing board for a game based on the nilpotent properties of the plane. In the game, which we just call $\prod_1$, the playing board consists of the regular $\prod_1$ hexagon H_7 illustrated in Figure 1.18.

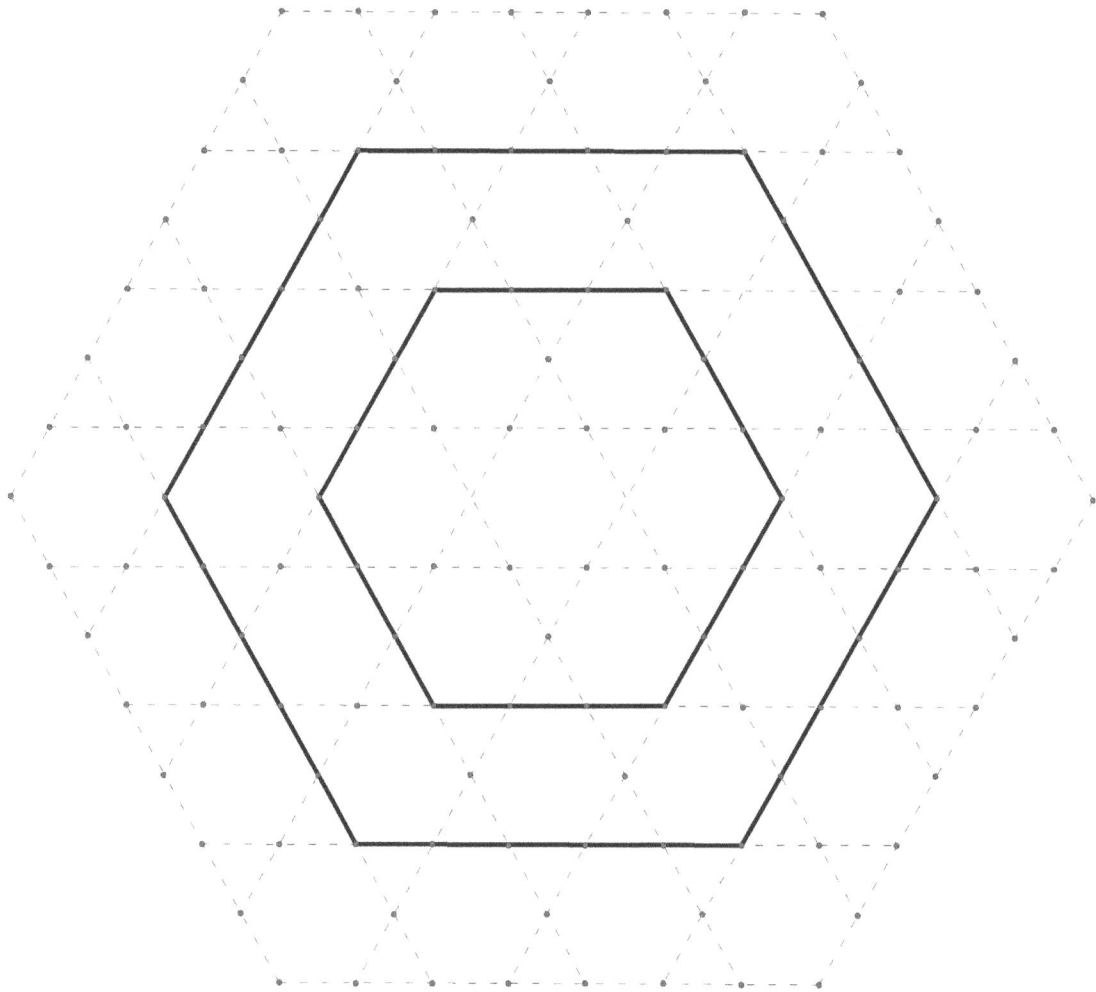

Figure 1.18

The playing board has 132 points and is accompanied by 130 playing pieces in the shape of small discs. Each disc is numbered on one side with the values {0,1,2,3,4,5,6,7,8,9}. In total there are 13 sets of ten discs with these numbers. Discs are played consecutively on points of the playing board and scores are achieved when a nilpotent $\prod_1$ polygon is completed with a total value which is a multiple of 5.

Discs are placed so that a connected subgraph of the playing board is maintained. There are two types of placements: (1) a *scoring position,* and (2) a *nonscoring position.* A scoring position is a point which completes a Δ_1 triangle or an H_1 hexagon with a multiple of 5. In such positions *only a scoring disc can be placed.* A nonscoring position is a point adjacent to a played disc but which does not complete a Δ_1 triangle or an H_1 hexagon.

The game probably works best with two players which we will call player A and player B. The totality of the playing discs are placed number side down on the table. This collection of unplayed tiles will be called the *boneyard*. Both players select 10 discs from the boneyard and place them face up before them in a rack, or some such contrivance, so they can be viewed only by the selecting player. The selected discs constitute the *hand* of each player. One more disc is selected from the boneyard and placed face up on one of the points of the central H_1 hexagon.

Players alternately take *turns*. A turn consists of making a series of scoring placements followed by a nonscoring placement. If no scoring placements are available, then a single nonscoring placement is made and the turn ends. *Only when a turn ends is a player allowed to replenish his/her hand to a full complement of 10 discs.* It is possible that a player can play all 10 tiles in one turn; in which case the turn also ends. It is also possible that a player can neither play a scoring or nonscoring disc. In this case a player *passes* without making any play, and thus ends his/her turn.

Say player A goes first. He places a disc from his hand adjacent to the central disc. No score is possible so one disc is drawn from the boneyard and play passes to B. Suppose the 6 in Figure 1.19(I) is the starting disc and A places the 2. Now B places the 7 to score 15 points, and finishes her hand by playing the nonscoring 3. A now places the 5 scoring 10 points and finishes his turn by placing the nonscoring 4, as I Figure 1.19(II).

B now places an 8 to score 15 points and finishes her turn by placing the nonscoring 1, as in Figure 1.19(III). A now places a 6 to score 15 points, and then places a 0 to score an H_1 worth 20 points. A continues his turn by placing a 9 scoring 10 points and then a 4 completing a J_1 for an additional score of 55 points. A thus scores a total of 100 points in this turn, and completes his turn by placing the nonscoring 1 as depicted in Figure 1.19(IV).

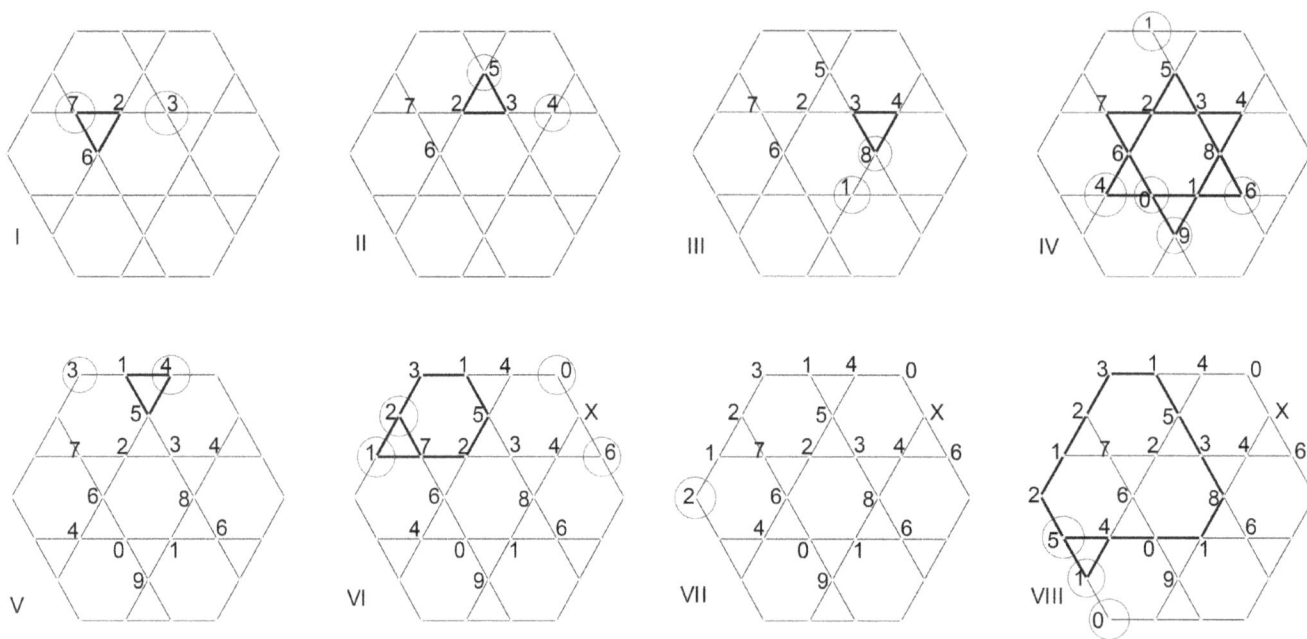

Figure 1.19

Play continues as in Figure 1.19(V) with B placing a 4 and then a 3 with a score of 10. Next A places a 2 to score an H_1 for 20 points, and then a 1 to score 10 points, and finally a nonscoring 0. B then has no scoring opportunities and so places a 6 in a nonscoring position. The point labeled X in Figure 1.19(VI) marks a *dead point*. It is dead since to complete the incident hexagon and triangle two different values would be required: a 0 or 5 for the triangle, and a 4 or 9 for the hexagon. No disc can be placed on a dead point, and no scoring configuration can contain a dead point in its interior or on its boundary. Players can create dead points in order to impede scoring opportunities for the opponent.

All plays so far have been confined to the central H_3 hexagon indicated in bold in Figure 1.18. This hexagon is a *barrier*. All points within or on the barrier must either be covered by discs or be dead before any play is allowed outside the boundary. This rule reduces the number of dead points and thus allows for the occurrence of larger scoring configurations.

Now suppose play continues from the position in Figure 1.19(VIII) until the position illustrated in Figure 1.20(I) occurs with B to play.

20

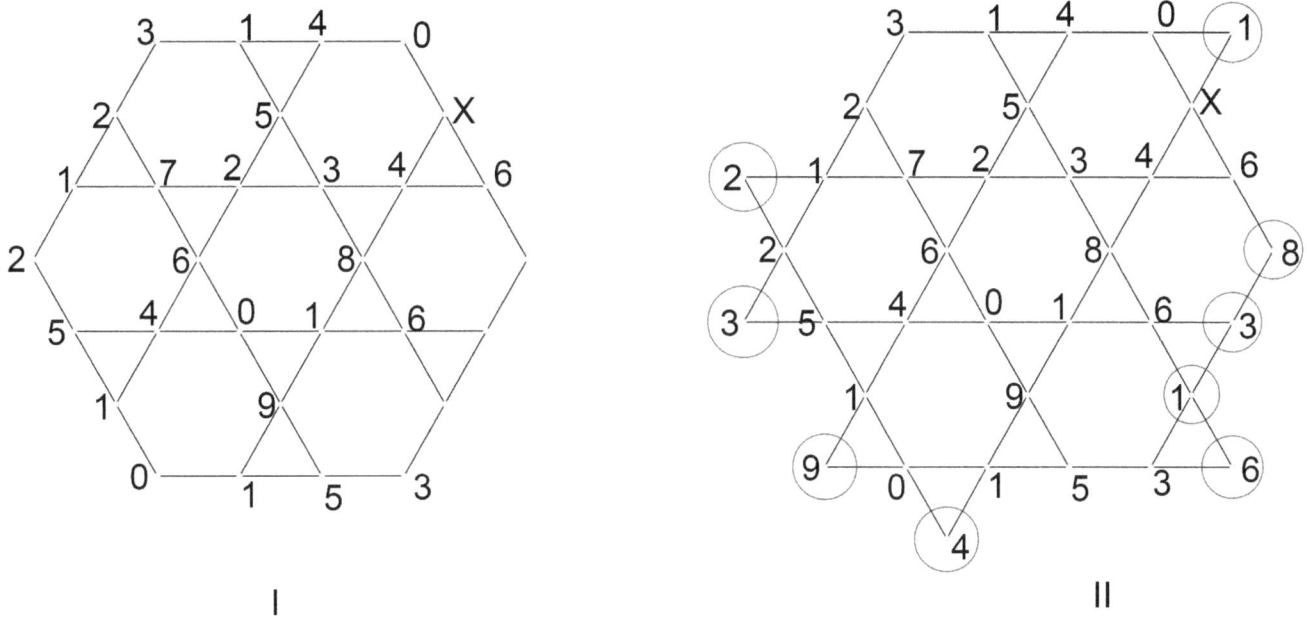

Figure 1.20

Now B plays scoring discs in the following order: (1) B plays 1 scoring an H_1 for 25 points; (2) 3 is played, adjacent to the 1 just played, scoring a Δ_1 for 10 points; (3) 6 is played scoring a Δ_3 for 40 points; (4) 9 is played completing a Δ_5 for 70 points; (5) 2 is played scoring a Δ_3 for 30 points; (6) 4 is played with a *double score*. A double score occurs when two scoring configurations are completed simultaneously, *and neither configuration is entirely contained in the other*. In this case the played 4 completes both a Δ_5 for 60 points, and a J_1 star for 45 points. Thus B earns a total of 105 points for just this play. (7) 3 is then played by B completing an S_3 star cluster for 90 points. (8) finally B completes her turn by playing the nonscoring 1. B scores a total of 370 points in this turn.

In Figure 1.18 a second (H_5) barrier is indicated. The rule for this barrier is the same as the H_3 barrier: no play is allowed beyond this barrier until all points between the two barriers have been covered. A barrier can be transgressed only in case of two successive passes. If one player cannot make a legitimate play inside a barrier, then he/she passes the play to the opponent. In the rare case that the opponent must also pass, then play reverts to the first player who is now allowed to play beyond the barrier.

There are many scoring configurations; the ones that occur most frequently are Δ_1, H_1, H_3, Δ_3, Δ_5, Δ_7, J_1, H(1,3,1,3,1,3), H(1,5,1,5,1,5), S_3. All these should certainly be included as scoring configurations. The players may also include conditionally nilpotent hexagons among the scoring configurations; the most commonly occurring ones are H(1,1,3,1,1,3) and H(1,3,3,1,3,3) as illustrated in Figure 1.16(I,II). The S_2 star cluster in Figure 1.17(I) can also be included. Composite star configurations which we denote by J^2 or J^3 do also occur occasionally, and these also might be included among the scoring configurations.

Be forewarned that the more scoring configurations are included, the more time consuming is the tabulation of the scores. There are, however, ways to accomplish score tabulation with more facility. For example, large triangles like Δ_7 can be partitioned into Δ_1 triangles for counting purposes, and H(1,5,1,5,1,5) can be scored by subtracting corner point values from a Δ_7.

Play continues until no more points can be legitimately covered. Players pass if they have no possible play, and when both players pass, the game ends and scores are tabulated. In case a shorter game is desired, the playing board can be designed as in Figure 1.21.

The rules are the same as in the larger version. There is just one H_3 boundary as indicated in Figure 1.21. There are only 90 points, so correspondingly 90 numbered discs consisting of 9 groups with values {0,1,2,3,4,5,6,7,8,9} should be used. We call this version the *game of stars* since star configurations such as J_1, S_3 and J^2 occur prominently as scoring configurations.

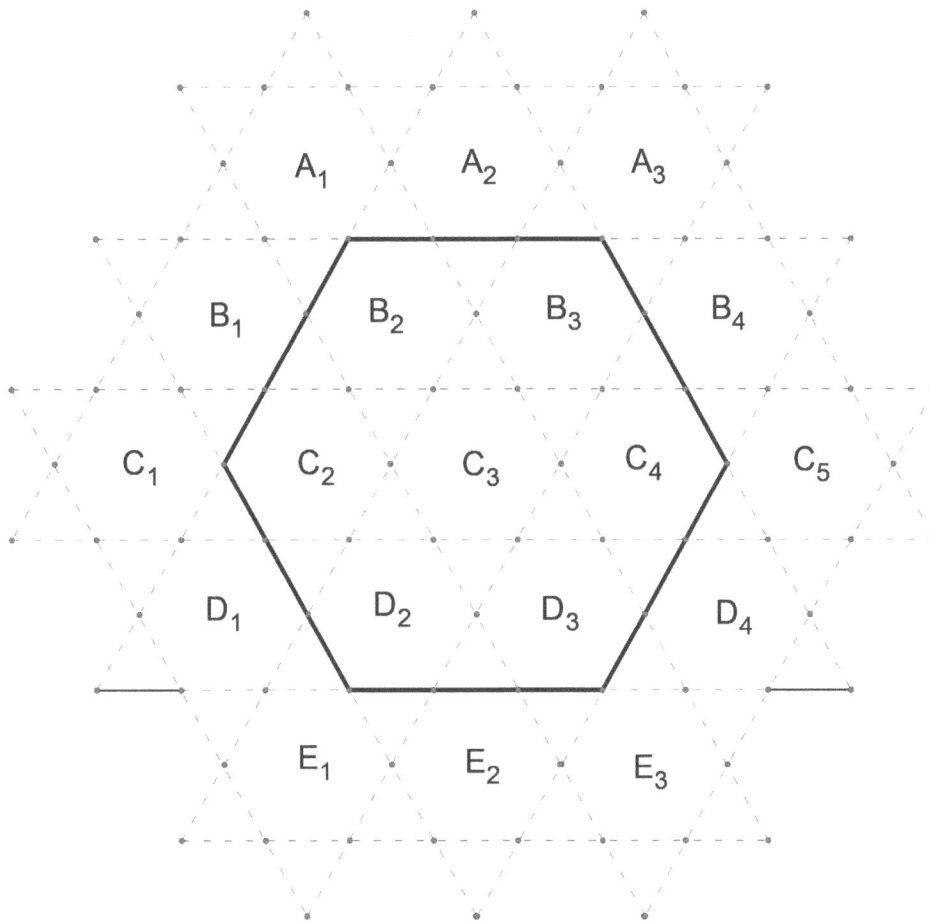

Figure 1.21

Blocking occurs when a nonscoring disc is played which creates an adjacent dead point. Disallowing blocking removes a basic strategy which can be used to limit the opponents scoring potential, but it makes more likely some of the larger scoring configurations. Blocking should be allowed in the larger version in order to limit the size of scoring configurations. For the smaller version blocking can be disallowed. In both versions playing a nonscoring disc in a region occupied by a dead point should be disallowed.

One more possible scoring configuration is illustrated in Figure 1.22. It consists of three Δ_1 triangles attached to the three vertices of a fourth Δ_1. We denote by Δ_1^4. In case blocking is not allowed, this configuration occurs so frequently that it seems preferable not to count it. Including Δ_1^4 as a scoring configuration is better suited to the game when blocking is allowed.

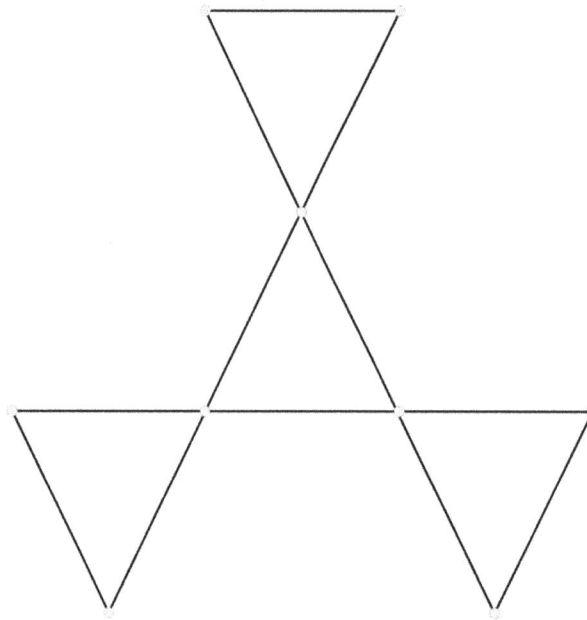

Figure 1.22

Also, with blocking disallowed, the H_5 barrier is not needed.

CHAPTER 2

Nilpotent structures in the hexagon tiling

2.1 Introduction

We will denote by $\prod_2$ the tiling of the plane by regular hexagons. Our purpose will be again to determine the nilpotent subgraphs of this plane. The *basic* nilpotent structures, the *tripod* and the hexagon H_1 are illustrated at left in Figure 2.1. In any locally nilpotent assignment these two figures must have 0 value.

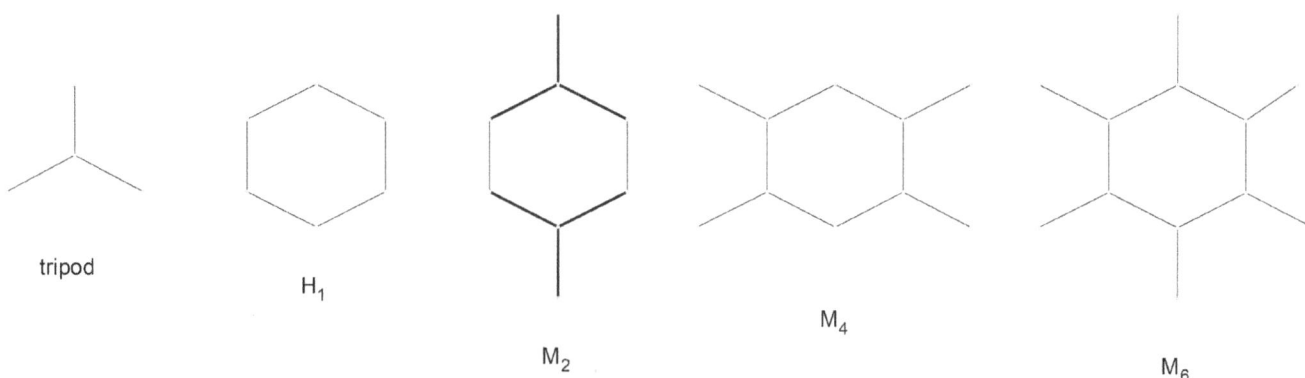

tripod H_1 M_2 M_4 M_6

Figure 2.1

The remaining structures in Figure 2.1 we call *mites* with 2, 4, 6 *point extensions*. Points which extend opposite points of an H_1 we call *point couples*. The mites are nilpotent as a consequence of the following simple observation.

Theorem 2.1 *In any locally nilpotent assignment, the values assigned to a point couple sum to 0.*

An application of Theorem 2.1 is illustrated in Figure 2.2.

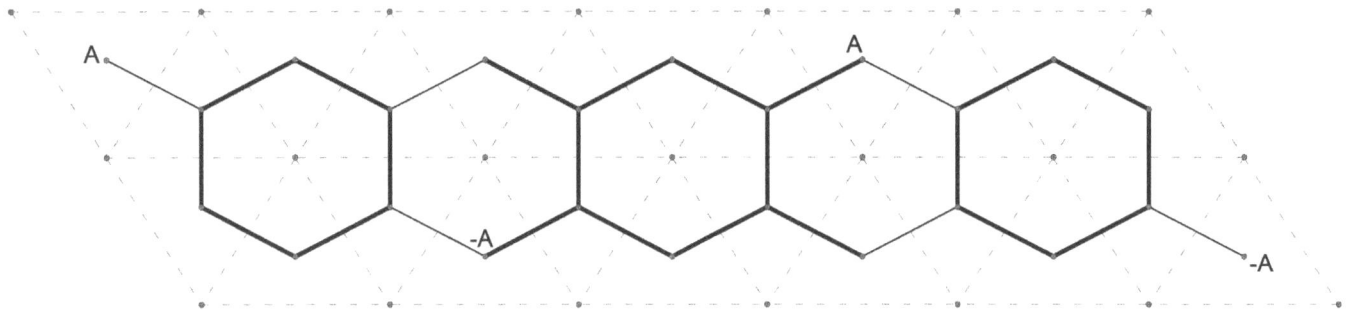

Figure 2.2

In Figure 2.2 a linear sequence of 5 hexagons, we denote by L_5, is illustrated. The indicated factorization of L_5 into two H_1's and an M_4 shows that L_5 is composite, (hence nilpotent). The structure remains nilpotent, but no longer composite, when the point extensions labeled A, -A are added. (In this Chapter the term "nilpotent" will mean "intrinsically nilpotent".) This labeling is justified by the indicated even length *chain* of point couples.

The dual of $\prod_2$ is the lattice of equilateral triangles which we will denote by Δ. A portion of this lattice is illustrated by dashed lines in Figure 2.2. Nilpotent configurations in the plane Δ are studied in [2], and due to duality, corresponding results are obtained for the hexagon lattice $\prod_2$. These results will be listed in the following Sections. The underlying triangle lattice also provides a way to represent nilpotent structures in $\prod_2$ as they are represented in [2]. L_5 can thus be given the hexagon label H(1,1,5,1,1,5), and with the two point extensions the parallelogram P(2,6) is obtained.

2.2 Triangles

Triangular configurations will be denoted by Δ_m, where the subscript m represents the number of unit edges per side in the underlying triangular lattice. In Figure 2.3 the triangles Δ_4 and Δ_6 are illustrated.

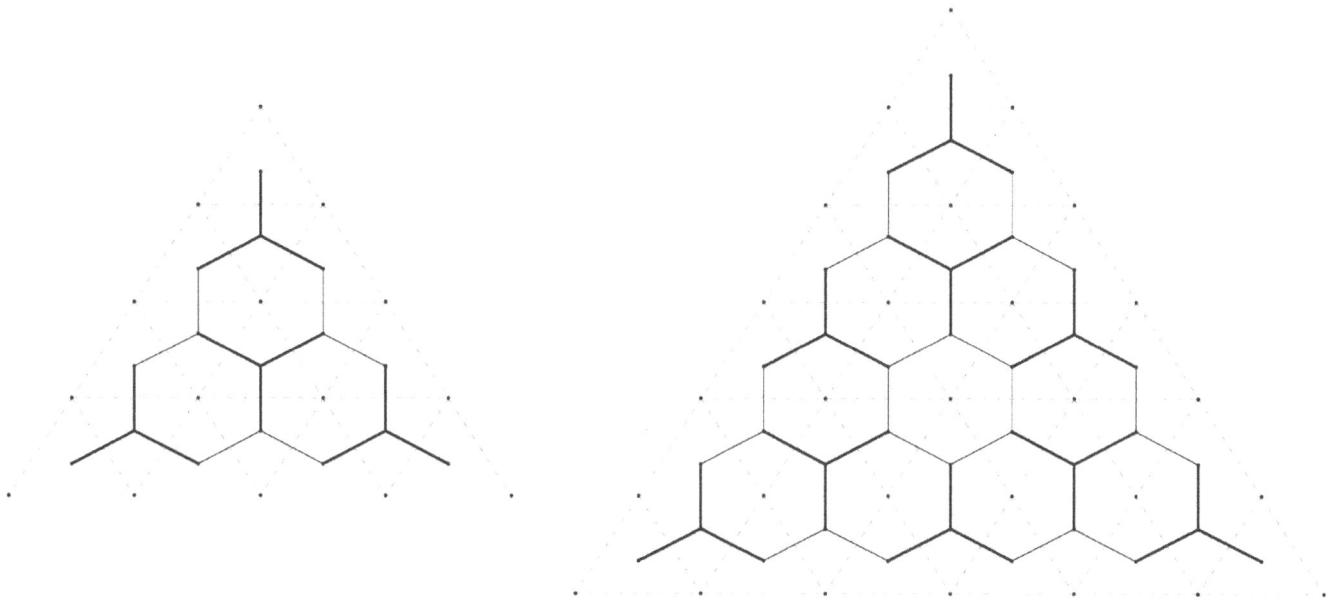

Figure 2.3

The triangles Δ_4, Δ_6 have tripod dissections, as indicated in Figure 2.3. In fact Δ_m has such a dissection whenever m is even.

Theorem 2.2 *The $\prod_2$ triangle Δ_m is composite, (hence nilpotent), iff m ≥ 4 is even.*

The "only if" part of Theorem 2.2 can be seen by considering the locally nilpotent assignment for Δ_5 illustrated in Figure 2.4. We call this assignment from {0,1,-1} a *linear pattern*. The value of this locally nilpotent assignment is 1, and consequently Δ_5 is not nilpotent.

2.3 Trapezoids

The trapezoids in $\prod_2$ are all isosceles, and will be denoted by T(a,b,a+b) where a, a+b represent the lengths of the bases in the underlying triangle lattice, and b represents the common length of the lateral sides.

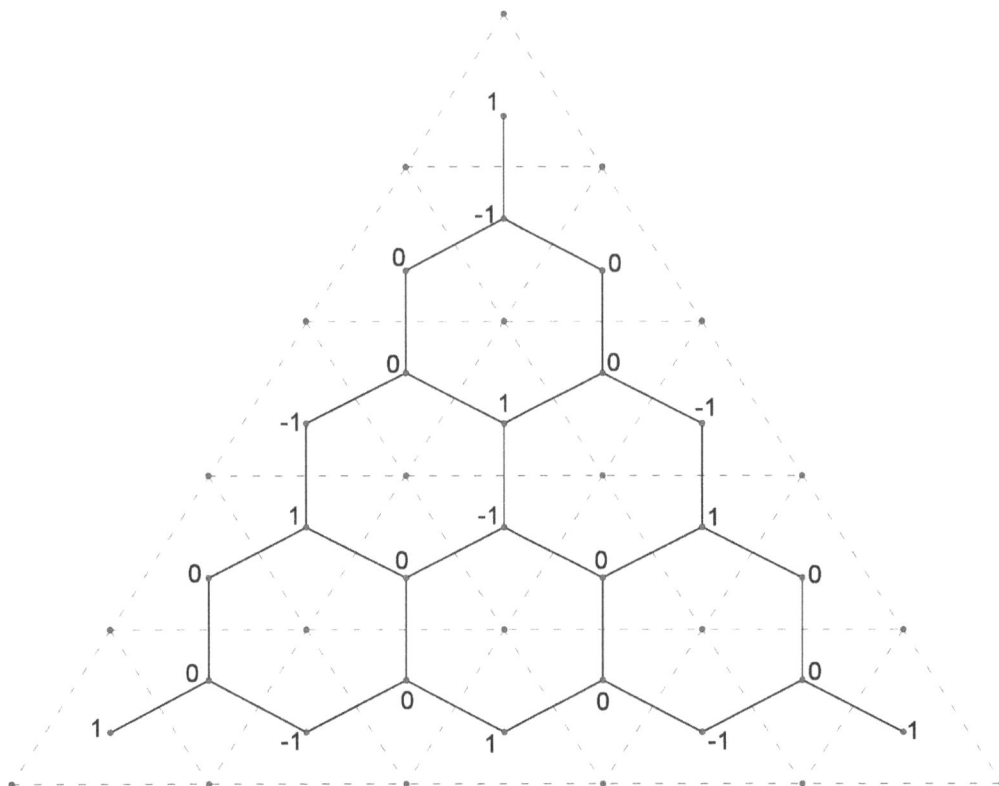

Figure 2.4

In case a, b are even then the trapezoid has a tripod dissection, as illustrated in Figure 2.5, and thus is nilpotent. Some trapezoids with odd length bases are also nilpotent, as indicated in the T(1,4,5) trapezoid in Figure 2.6(II).

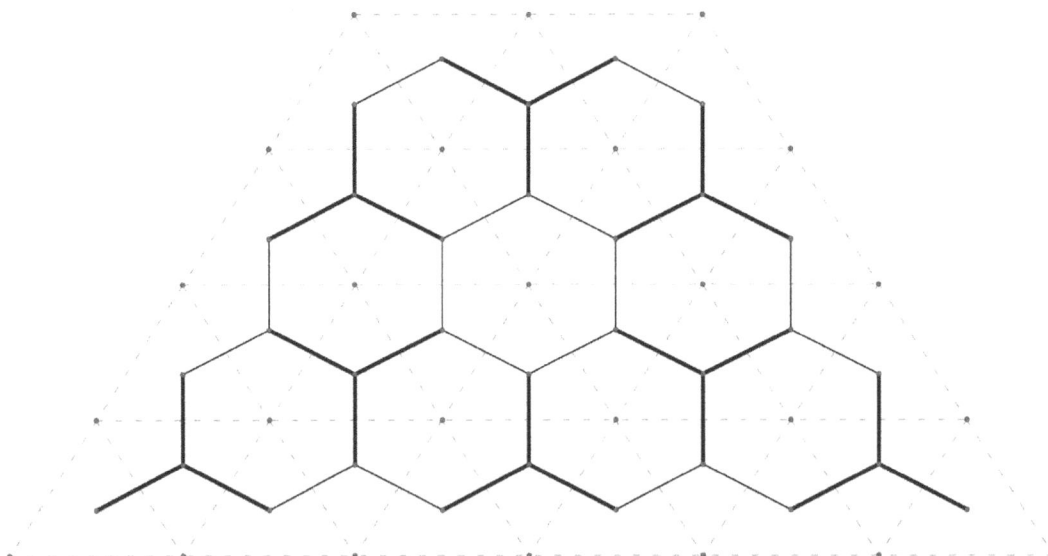

Figure 2.5

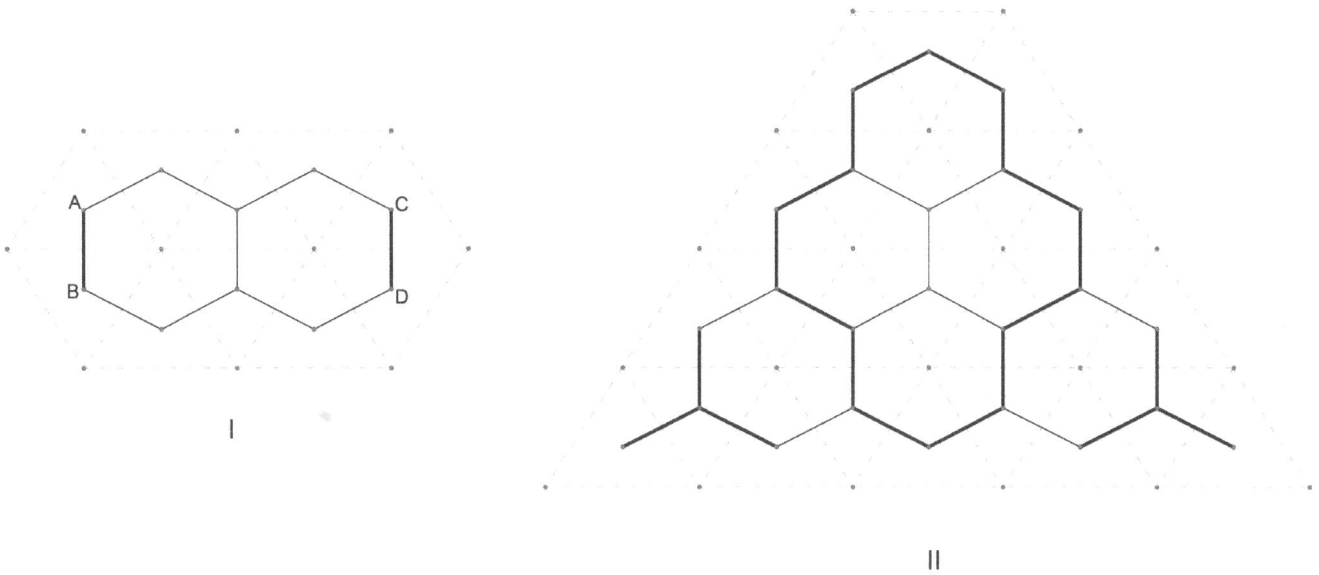

Figure 2.6

The nilpotence of the T(1,4,5) trapezoid in Figure 2.6(II) is dependent upon the observation illustrated in Figure 2.6(I).

Lemma 2.3 *In any locally nilpotent assignment the sum of the values on the opposite edges of a H(1,1,2,1,1,2) hexagon, as in Figure 2.6(I), is 0.*

We call the opposite edges, (A,B) and (C,D), in Figure 2.6(I), an *edge couple*. As a consequence the H(1,2,2,1,2,2), which occurs at the center of the T(1,4,5) in Figure 2.6(II), is nilpotent. As a further consequence the trapezoid T(1,4,5) is also nilpotent. More generally we have:

Theorem 2.4 *Every isosceles trapezoid with lateral sides a multiple of 4, is nilpotent.*

Even more generally, the following result can be obtained.

Theorem 2.5 *An isosceles trapezoid is nilpotent iff (i) all sides are even, or (ii) its lateral sides have length a multiple of 4.*

2.4 Parallelograms

The parallelogram P(a,b) has a Δ_2 dissection in case a, b are even. There also exist nilpotent parallelograms in which one of the sides has odd length. The parallelogram

29

P(4,3) is illustrated in Figure 2.7. The dissection into two tripods and an H(1,2,2,1,2,2) shows that P(4,3) is nilpotent.

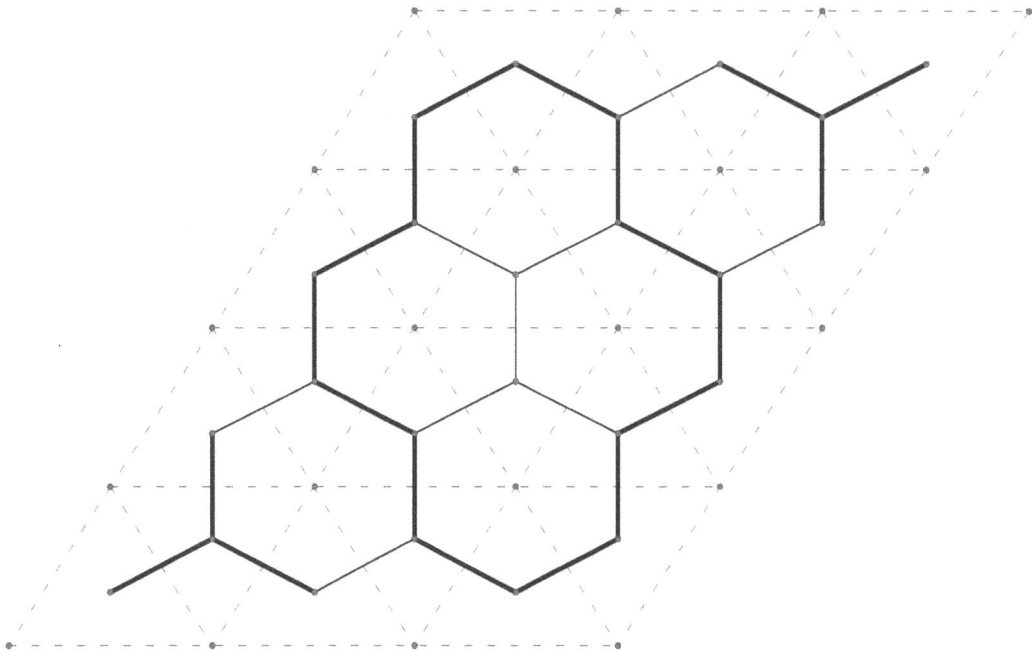

Figure 2.7

Now via extensions of the nilpotent P(4,3) the following result is obtained.

Theorem 2.6 *The parallelogram P(4t, q) is nilpotent for every positive integer t, and for every integer q ≥2.*

The linear pattern is again useful in proving the following Theorem.

Theorem 2.7 *If a, b are both odd then P(a,b) is not nilpotent.*

The final result concerning parallelograms is the following.

Theorem 2.8 *The parallelogram P(a,b) is nilpotent iff (i) a, b are both even, or (ii) one of the parameters, say a, is a multiple of 4, and b is any integer larger than or equal to 2.*

To complete the proof of Theorem 2.8 it remains only to show that P(4t+2,q) is nonnilpotent for odd q and any positive integer t. To this end the new pattern illustrated in Figure 2.8 is useful.

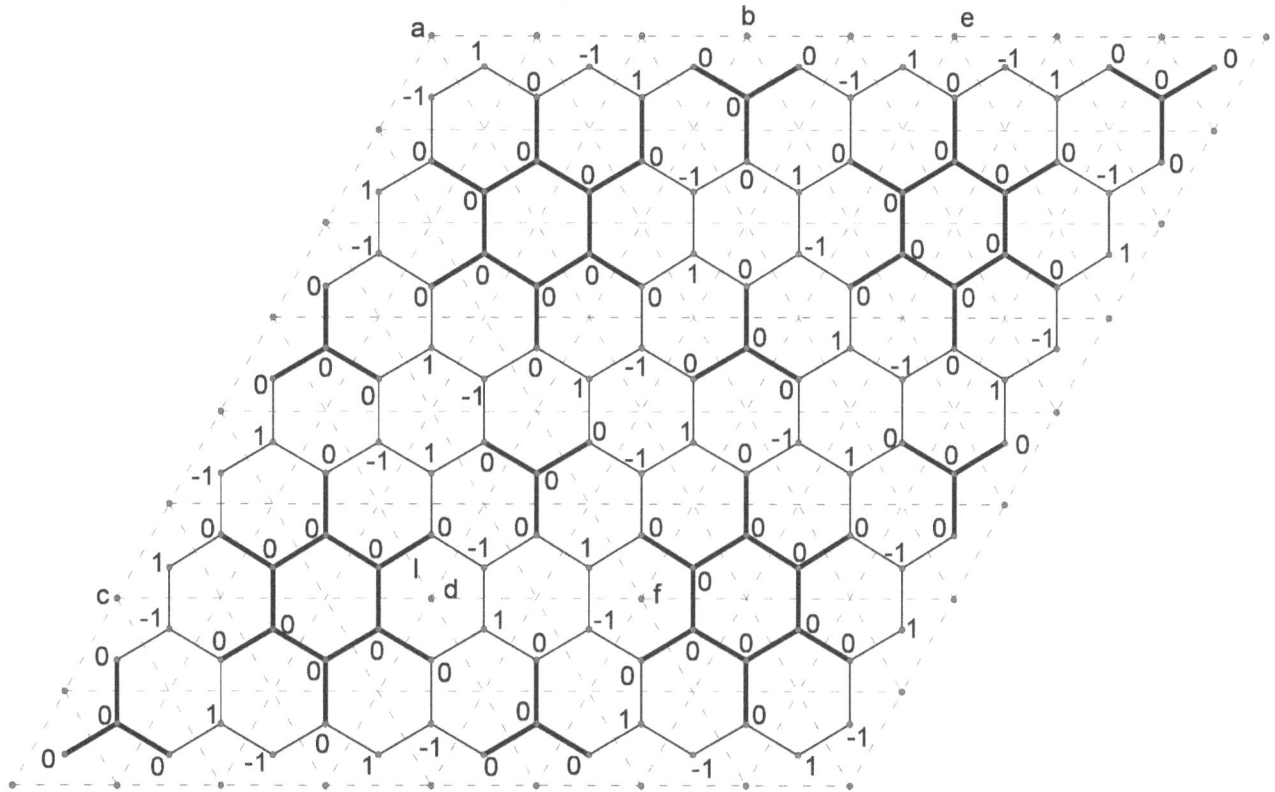

Figure 2.8

The locally nilpotent assignment in Figure 2.8 extends to the entire Π_2 plane, and we call it the *tripod and M_6 pattern* in accordance with the formations obtained by the 0 valued points. If we consider the P(6,3) parallelogram with corners {a,b,c,d}, we find that its value is 1. The conclusion is that P(6,3) is not nilpotent. Various extensions can be made consisting of parallelograms with all even sides. Since these all have value 0 under any locally nilpotent assignment, the desired result is obtained.

2.5 Regular and Semiregular Hexagons

We use H_m to denote the regular hexagon with m unit edges per side. In case m is even, then H_m has a tripod dissection and thus is nilpotent. In case m is odd, m≥3, a dissection can always be found using the prime polygons {Δ_2, H_1, M_4, M_6, H(1,2,2,1,2,2)}. In Figure 2.9 such a dissection is illustrated for H_3. In case m≥5 a large variety of prime factorizations of H_m can be constructed. The reader may try to construct some of these. We have the following result. (Note that Δ_2 is another name for a tripod.)

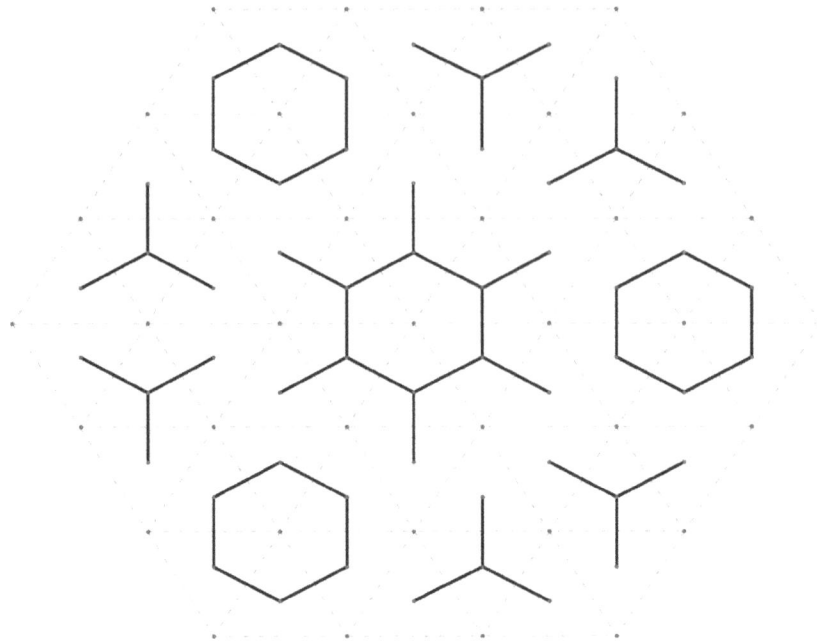

Figure 2.9

Theorem 2.9 *The regular hexagon H_m is nilpotent for every positive integer m.*

A *near -regular hexagon* has the form H(a,b,b,a,b,b). In case a = 1 and b is even, then a factorization involving a row of H(1,2,2,1,2,2) hexagons and a collection of tripods can be constructed. Such a dissection is illustrated for H(1,4,4,1,4,4) in Figure 2.10.

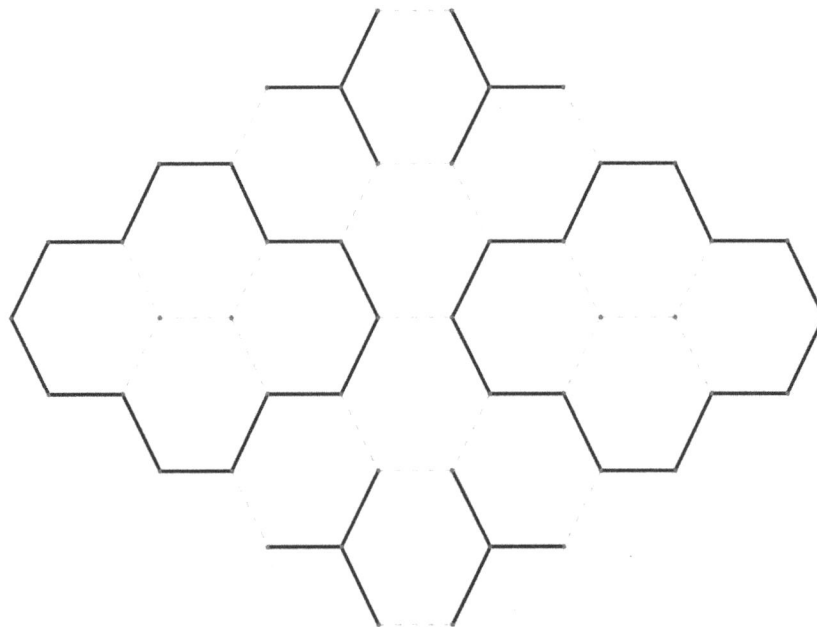

Figure 2.10

Hexagons of the form H(1,b,b,1,b,b) where b is odd with the form b = 4t+3 can be shown to be nonnilpotent using the linear pattern as a locally nilpotent assignment. In case b is odd with b = 4t+1, then H(1,b,b,1,b,b) has a factorization consisting of the parallelogram P(4t,4t) encased on one side by two copies of a nilpotent H(1,1,4t+1,1,1,4t+1) which have an H_1 in common. This dissection is illustrated in an H(1,5,5,1,5,5) hexagon in Figure 2.11. As a consequence we have the following Theorem.

Theorem 2.10 *The near-regular hexagon H(1,b,b,1,b,b) is nilpotent except when b has the form b = 4t+3.*

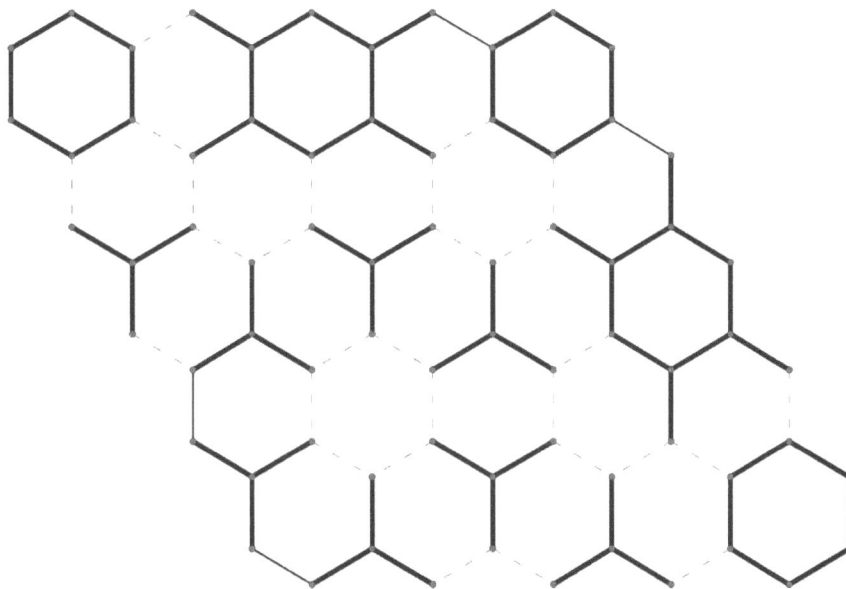

Figure 2.11

A *semiregular hexagon* has the form H(p,q,r,p,q,r). In case the parameters p,q,r are all even, then the hexagon has a tripod dissection, and consequently is nilpotent. Suppose p is even and q, r are odd. In this case H(p,q,r,p,q,r) can be extended to the parallelogram P(q+p, r+p) by appending two copies of the triangle Δ_p. The parallelogram P(q+p, r+p) is nonnilpotent by Theorem 2.7, a fact which can be shown using the linear pattern. Since p is even, the extension of this linear pattern yields a 0 value for each of the appended triangles. As a consequence H(p,q,r,p,q,r) remains nonnilpotent when the appended triangles are removed.

Now suppose p, q are even; r is odd and p+q is a multiple of 4. Extend H(p,q,r,p,q,r) again by appending two Δ_p triangles to the opposites of length p. The result is the parallelogram P(p+q, r) which is nilpotent by Theorem 2.6. The hexagon H(p,q,r,p,q,r) remains nilpotent when the appended nilpotent triangles are removed. In Figure 2.12 the hexagon H(2,2,5,2,2,5) is illustrated with a prime factorization consisting of an H(1,2,2,1,2,2) and several tripods.

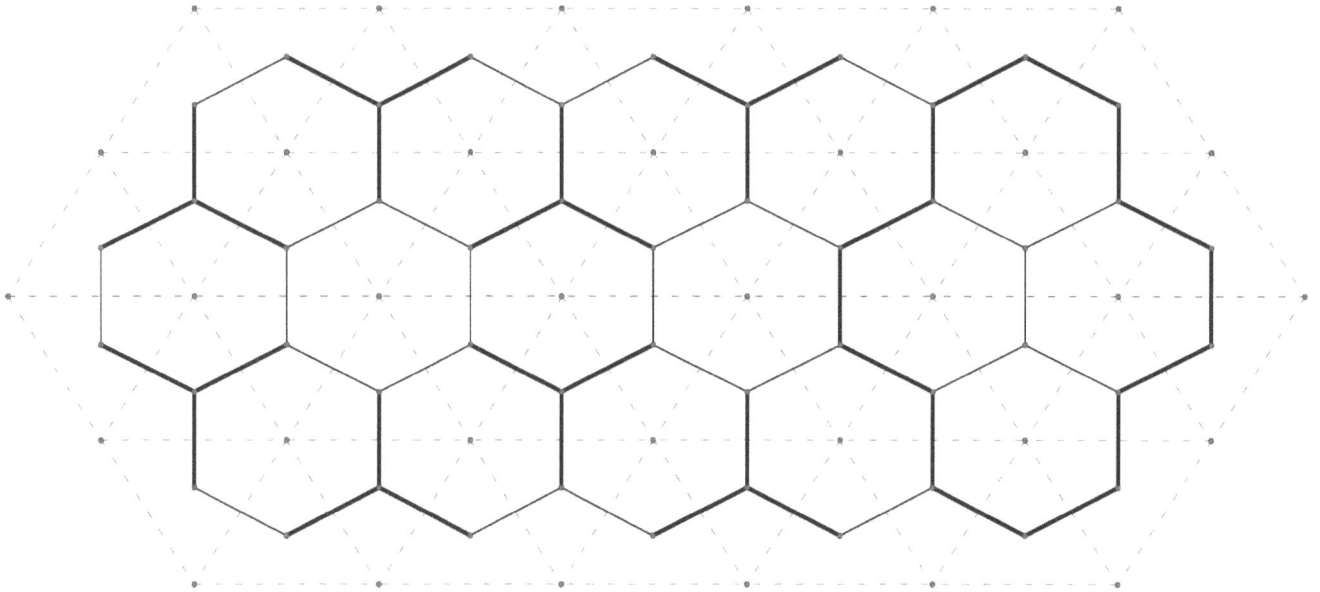

Figure 2.12

In case p, q are even; r is odd and p+q is not a multiple of 4, Then the parallelogram P(p+q, r) obtained by extension is not nilpotent by Theorem 2.8, and it then follows that H(p,q,r,p,q,r) is also nonnilpotent. We thus have the following result.

Theorem 2.11 *If p, q are even and r is odd, then the semiregular hexagon H = H(p,q,r,p,q,r) is nilpotent iff p+q is a multiple of 4. In case one of the parameters is even and the remaining two parameters are odd, then H is nonnilpotent.*

It remains to consider semiregular hexagons H(p,q,r,p,q,r) in which all parameters are odd. In Figure 2.13 the extension of H(p,q,r,p,q,r) by the nilpotent parallelograms P(r,4s) and P(q,4s) is illustrated. As a consequence, H(p+4s, q, r, p+4s,q,r)) is nilpotent iff H(p,q,r,p,q,r) is. The nilpotence of H(p,q,r,p,q,r) thus depends upon the nilpotence of H_1= H(1,1,1,1,1,1), H(1,1,3,1,1,3), H_3 = H(3,3,3,3,3,3), or H(1,3,3,1,3,3). The simple hexagon H_1 is nilpotent by definition and H_3 is composite hence nilpotent, as in Figure 2.9. On the other hand H(1,3,3,1,3,3 is nonnilpotent by Theorem 2.10, and H(1,1,3,1,1,3) is easily seen to be nonnilpotent. Thus any nilpotent extension of these two hexagons is also nonnilpotent. As a consequence we have the following Theorem.

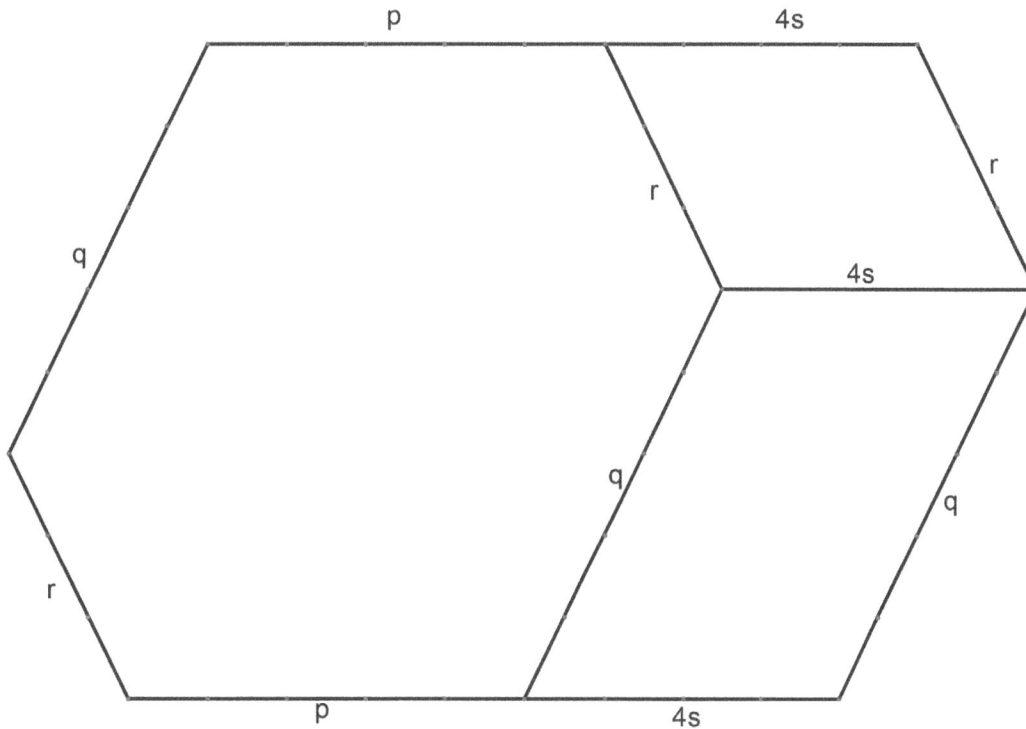

Figure 2.13

Theorem 2.12 *Let H(p,q,r,p,q,r) be a semiregular hexagon with p,q,r all odd. Then H(p,q,r,p,q,r) is nilpotent iff (i) each of p,q,r is congruent to 1 mod 4, or (ii) each of p,q,r is congruent to 3 mod 4.*

In Figure 2.14 the semiregular hexagon H(1,5,5,1,5,5) is illustrated with a prime factorization consisting of H_1 hexagons, tripods and M_2 mites. In Figure 2.15 the semiregular hexagon H(3,3,7,3,3,7) is illustrated with a prime factorization consisting of H_1 hexagons, tripods, an M_6 mite, and two H(1,2,2,1,2,2) hexagons.

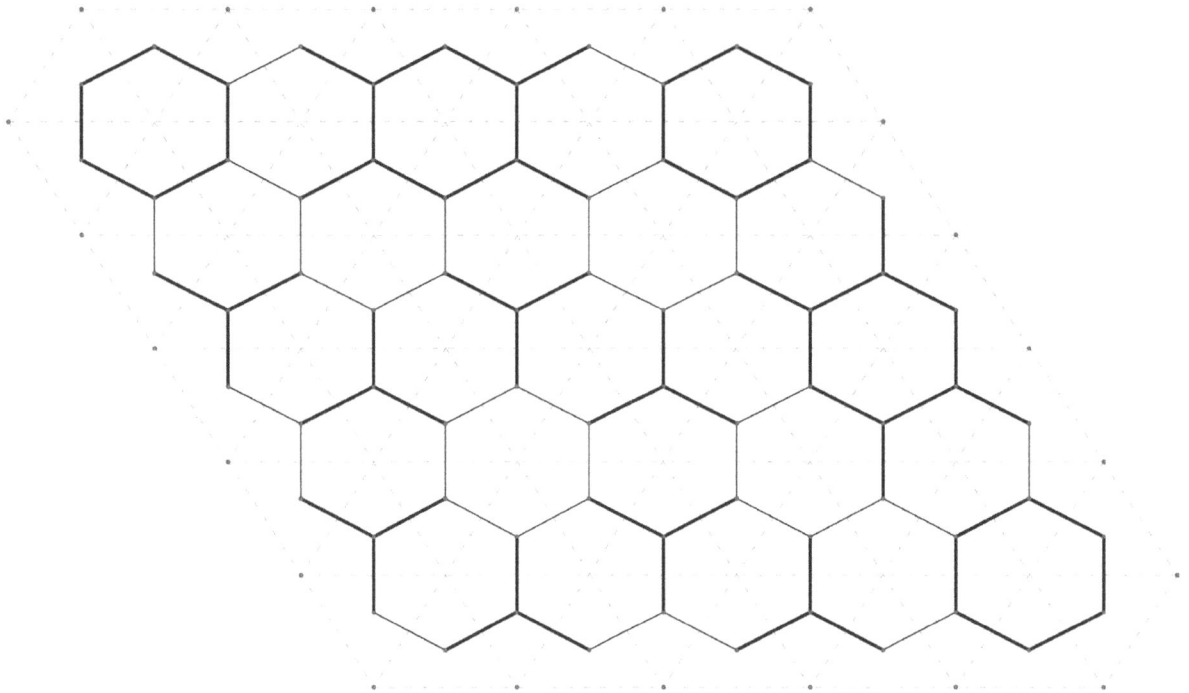

Figure 2.14

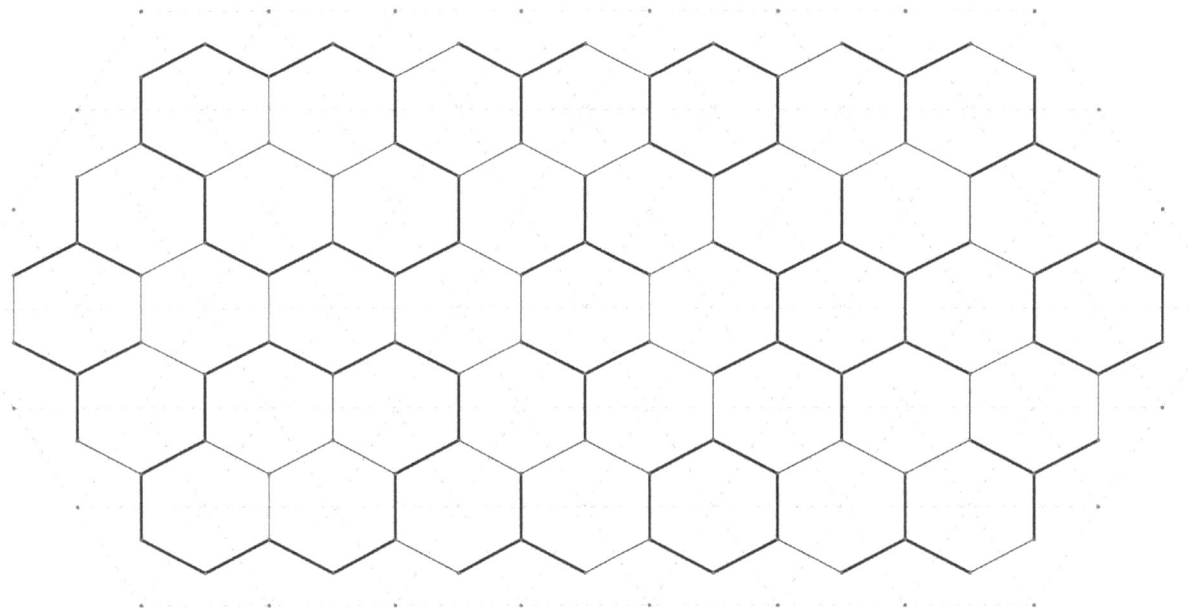

Figure 2.15

2.6 Alternating Hexagons

In this Section we consider nonsemiregular hexagons of the form H(a,b,a,b,a,b) with a ≠ b. These we call *alternating*. Suppose a, b are odd with b = a+4 as in Figure 2.16.

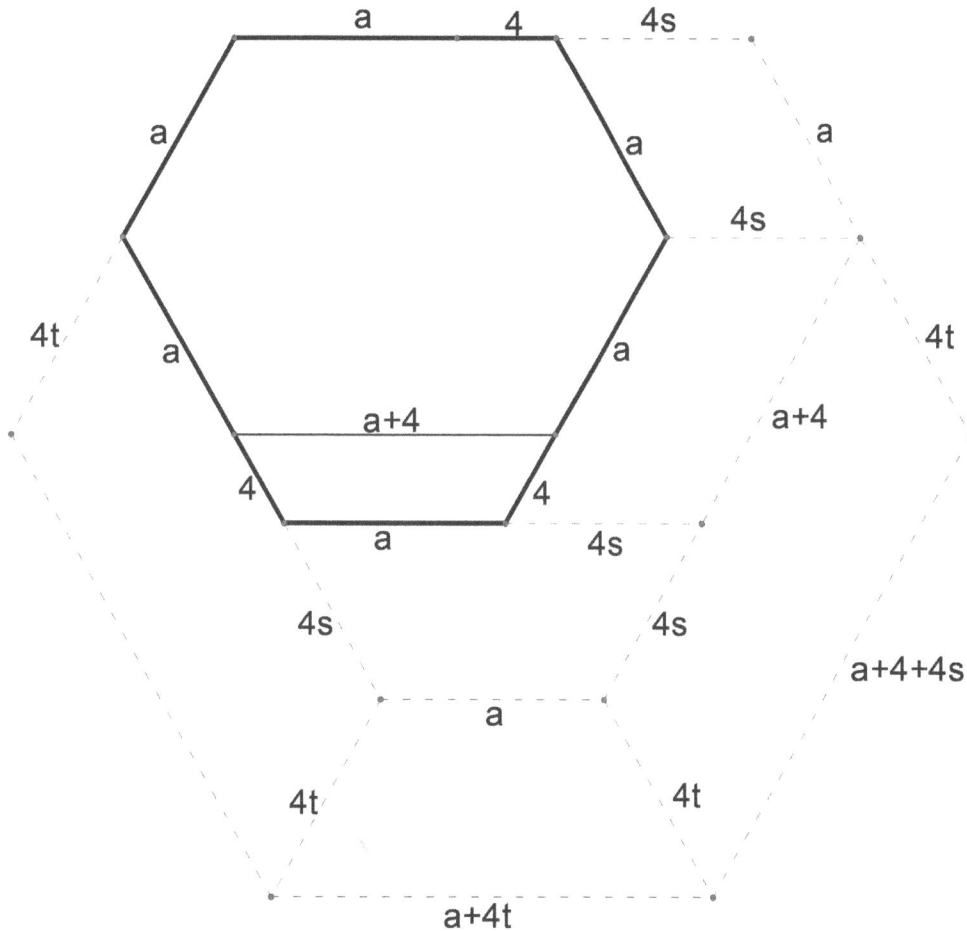

Figure 2.16

In Figure 2.16 the alternating hexagon H(a, a+4, a, a+4, a, a+4) is outlined in bold. It is partitioned into a semiregular hexagon H(a, a, a+4, a, a, a+4) and the isosceles trapezoid T(a, 4, a+4). The hexagon is nilpotent by Theorem 2.12, and the trapezoid is nilpotent by Theorem 2.4. As a consequence the indicated alternating hexagon is also nilpotent.

Also in Figure 2.16 the extension sequence H(a, a+4, a, a+4, a, a+4) $\rightarrow$ H(a, a+4+4s, a, a+4+4s, a, a+4+4s) $\rightarrow$ H(a+4t, a+4+4s, a+4t, a+4+4s, a+4t, a+4+4s) is accomplished by the annexation of nilpotent parallelograms and trapezoids. As a consequence the alternating hexagon H(p,q,p,q,p,q) is nilpotent whenever p, q have the form p = a+4t, and q = a + 4s with a odd. This is equivalent to saying p+q $\equiv$ 2 mod 4. More precisely the following Theorem holds.

Theorem 2.13 *The alternating hexagon H(p,q,p,q,p,q) with p, q odd, is nilpotent iff p+q ≡ 2 mod 4.*

The proof of the "only if" part of Theorem 2.13 can be found in [2]. The general hexagon H(p,q,r,s,t,u) with all sides of even length, has a tripod dissection and thus is nilpotent. In particular, when both a, b are even, the alternating hexagon H(a,b,a,b,a,b) is nilpotent. It thus remains only to consider alternating hexagons with a even and b odd. In [2] it is shown that none of these are nilpotent. Here we will just consider the case with a = 4s+2, and b with the form b = 4t+3. Figure 2.17 illustrates the general structure of such a hexagon.

In this Figure the trapezoids T_1, T_2 are nilpotent by Theorem 2.4, and the parallelograms P_1, P_2, P_3 are nilpotent by Theorem 2.8. The region labeled H is the alternating hexagon H(2,3,2,3,2,3), and the region labeled Q^t represents t copies of the region labeled Q in Figure 2.18.

Figure 2.18 is a blow-up of the region in bold in Figure 2.17, (with t = 1). Except for the 5 points labeled with 0, 1, -1, this region has the indicated prime factorization. The labeled points represent a part of a linear pattern which clearly extends to cover the entire region. Since the five points have a nonzero sum, it can be concluded that the H(2,3,2,3,2,3) hexagon as well as its extension to the entire H(4s+2, 4t+3, 4s+2, 4t+3, 42+2, 4t+3) hexagon is nonnilpotent. Theorem 2.13 can thus be strengthened as follows.

Theorem 2.14 *The alternating hexagon H(p,q,p,q,p,q) is nilpotent iff p, q are odd and p+q ≡ 2 mod 4.*

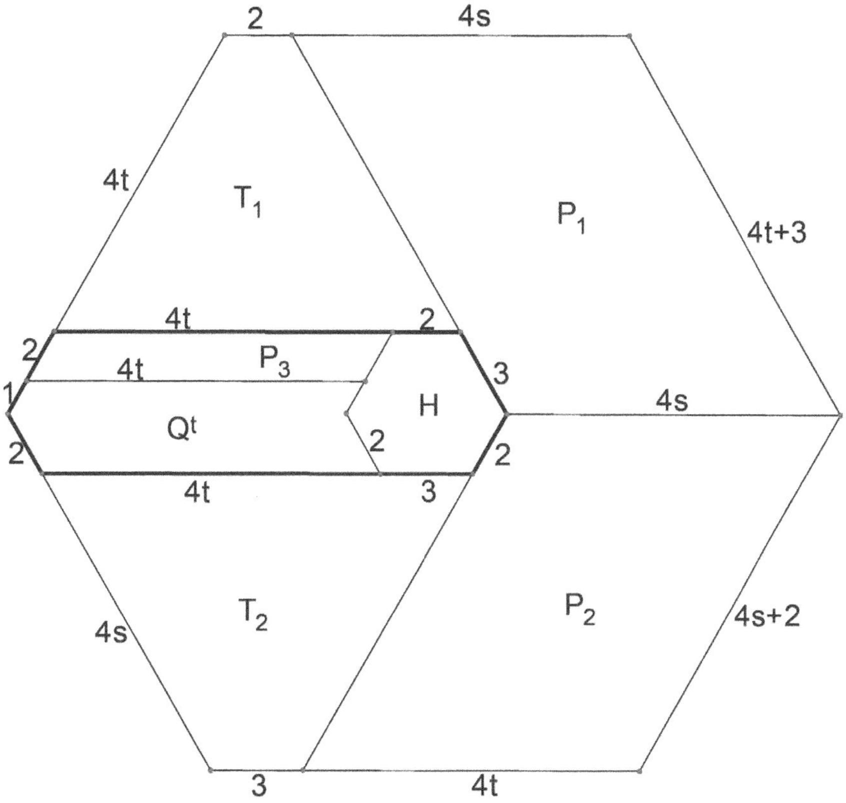

Figure 2.17

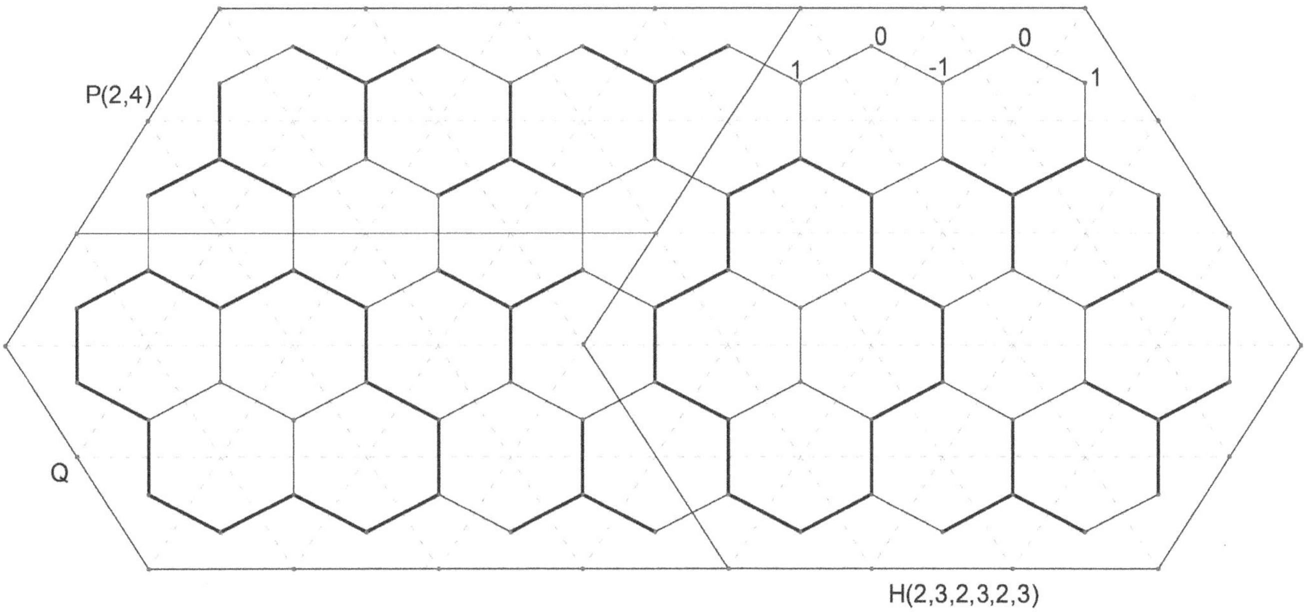

Figure 2.18

Before leaving the subject of alternating hexagons we look at a factorization of H(1,5,1,5,1,5) which reveals an important point. In Figure 2.19 a prime factorization of H(1,5,1,5,1,5) using H_1, M_4, some tripods and H(1,2,2,1,2,2). In Figure 2.20 a second factorization is given which consists of three hexagons in the corners and an odd shaped construction consisting of an H(1,3,1,3,1,3) hexagon with several point extensions. As embedded in the surrounding H(1,5,1,5,1,5) hexagon, it must have a value of 0 in any locally nilpotent assignment, *but to determine if it is actually nilpotent, it must be analyzed after extracting it from any surroundings.*

In Figure 2.21 the figure in question appears by itself and has been labeled for analysis.

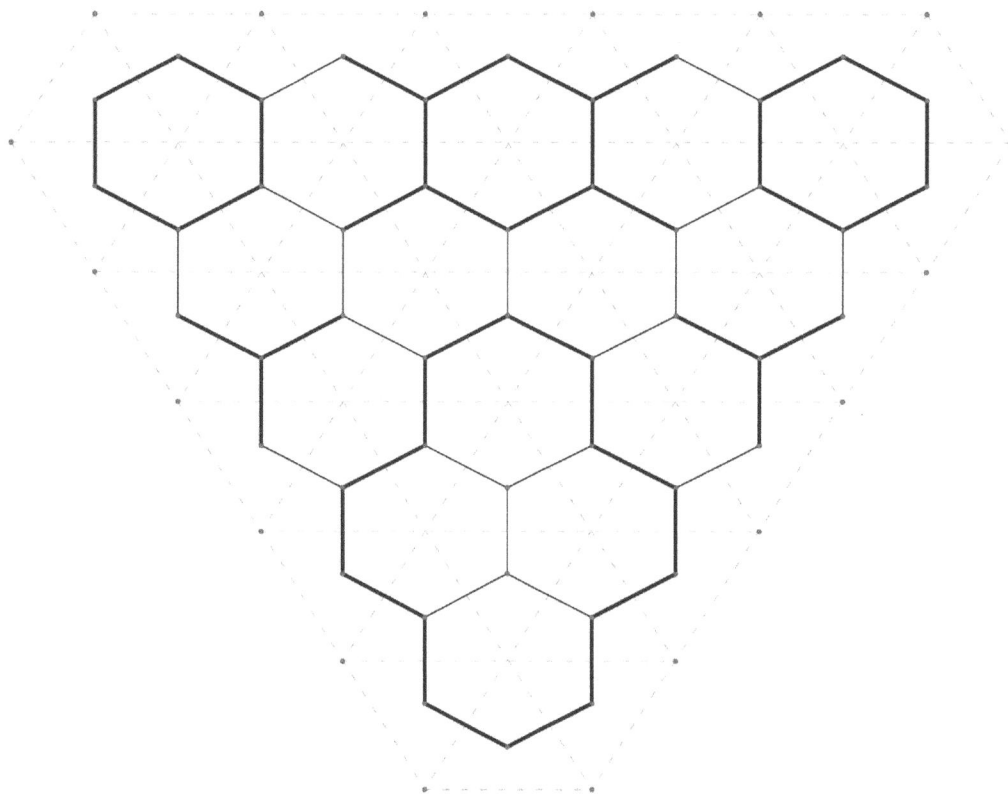

Figure 2.19

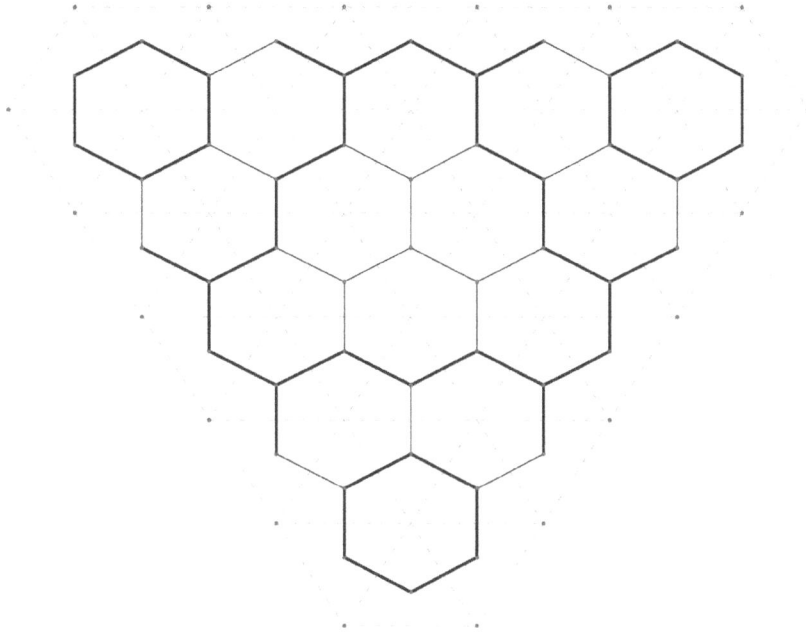

Figure 2.20

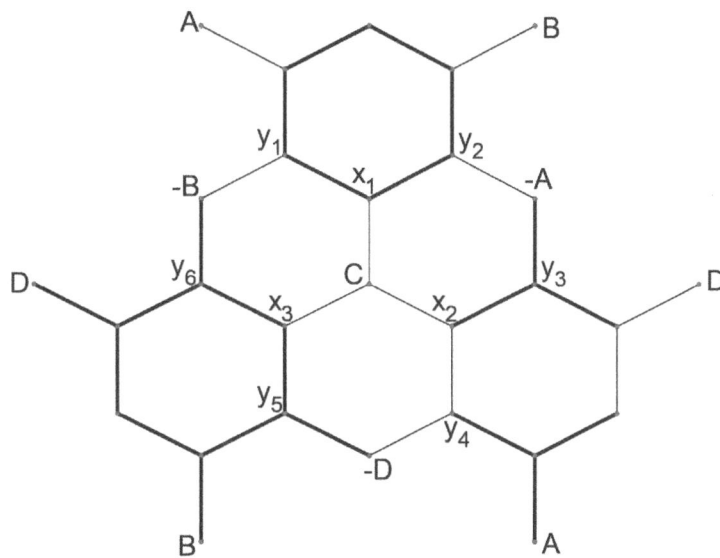

Figure 2.21

In accordance with the dissection indicated in Figure 2.21, to show the nilpotence of the figure, it suffices to show that A+B+C+D = 0. By considering the various tripods and hexagons which contain point C, we obtain the following equations.

(1) $C + x_1 + x_2 + x_3 = 0$

(2) $C + x_1 + y_1 + y_2 = 0$

(3) $C + x_2 + y_3 + y_4 = 0$

41

(4) $C + x_3 + y_5 + y_6 = 0$

(5) $C + x_1 + y_1 + x_3 + y_6 = B$

(6) $C + x_1 + x_2 + y_2 + y_3 = A$

(7) $C + x_2 + x_3 + y_4 + y_5 = D$

Equations (1), (2) yield

(8) $x_2 + x_3 = y_1 + y_2$;

equations (1), (3) yield

(9) $x_1 + x_3 = y_3 + y_4$,

and equations (1), (4) yield

(10) $x_1 + x_2 = y_5 + y_6$.

Now from (4), (5) we obtain

(11) $x_1 + y_1 - y_5 = B$;

from (4), (6), (10) we have

(12) $y_2 + y_3 - x_3 = A$,

and from (4), (7) we obtain

(13) $x_2 + y_4 - y_6 = D$.

Solving (4) for C, and then adding equations (11), (12), (13) yields

(14) $A + B + C + D =$

$$(y_2 + y_3 - x_3) + (x_1 + y_1 - y_5) + (-x_3 - y_5 - y_6) + (x_2 + y_4 - y_6) =$$

$$(y_2 + (y_3 + y_4) + (x_1 + x_2 - y_5 - y_6) - x_3 + (y_1 - x_3) + (-y_5 - y_6) =$$

$$(y_2 + (x_1 + x_3) + (0) - x_3 + (x_2 - y_2) + (-x_1 - x_2) = 0.$$

As a consequence of this demonstration, we can conclude that the graph in Figure 2.21 is nilpotent. Let us denote this figure by G_6, where $G = H(1,3,1,3,1,3)$ and the subscript represents a point extension of G with 6 points. No further dissection of G_6 into smaller nilpotent subgraphs is possible, so the factorization illustrated in Figure 2.20 is a prime factorization. We may attempt to factorize G_6 as illustrated in Figure 2.22.

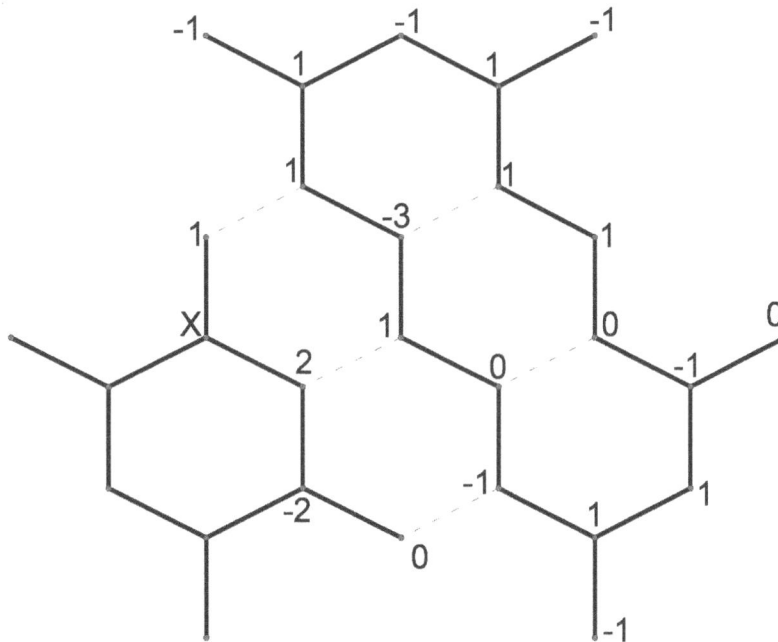

Figure 2.22

The factorization consists of an M_4 and a linear polyhex L_3 with 4 point extensions. To this last factor a locally nilpotent assignment has been provided. The sum of this assignment is -1, showing that L_3 with the four point extensions is not nilpotent. Any attempt to extend this assignment to a locally nilpotent assignment of G_6 is necessarily doomed to failure. In Figure 2.22 a *conflict* occurs at point X; to complete the tripod X = -1 is needed, but to complete the H_1, X = -2 is required. We can say that L_3 with the point extensions is *contextually nilpotent;* i.e., it is nilpotent only in some extension.

In any case, the above analysis tells something about any locally nilpotent assignment for the entire Π_2 plane: any four points A,B,C,D which occupy the four corners of a parallelogram, as in Figure 2.21 must sum to 0. We say the four points form a *nilpotent quadruple* in Π_2.

2.7 Irregular hexagons

For the general Π_2 hexagon H(p,q,r,s,t,u) the parameters must satisfy the equations

(1)   p+q = s+t,

(2)   r+q = t+u.

The proof of the following Theorem appears in [2].

Theorem 2.15 *Let H = H(p,q,r,s,t,u) be a Π_2 hexagon with all parameters odd. Then H is nilpotent iff (i) all parameters are congruent to 1 mod 4, or (ii) all parameters are congruent to 3 mod 4.*

The idea behind the proof of Theorem 2.15 has to do with a procedure we call *trapezoid reduction.* As an example of this process, consider the H(5,5,5,9,1,9) hexagon in Figure 2.23 in which all parameters are congruent to 1 mod 4. The sequence of reductions is given by H(5,5,5,919) $\rightarrow$ H(1,5,9,1,5,9) $\rightarrow$ H(5,1,9,1,5,5) $\rightarrow$ H(5,1,5,5,1,5) $\rightarrow$ H(1,5,1,5,1,5).

The last member of this sequence is a nilpotent alternating hexagon, and since the remaining part of H(5,5,5,9,1,9) is dissected into nilpotent trapezoids, we can conclude that the original hexagon is nilpotent.

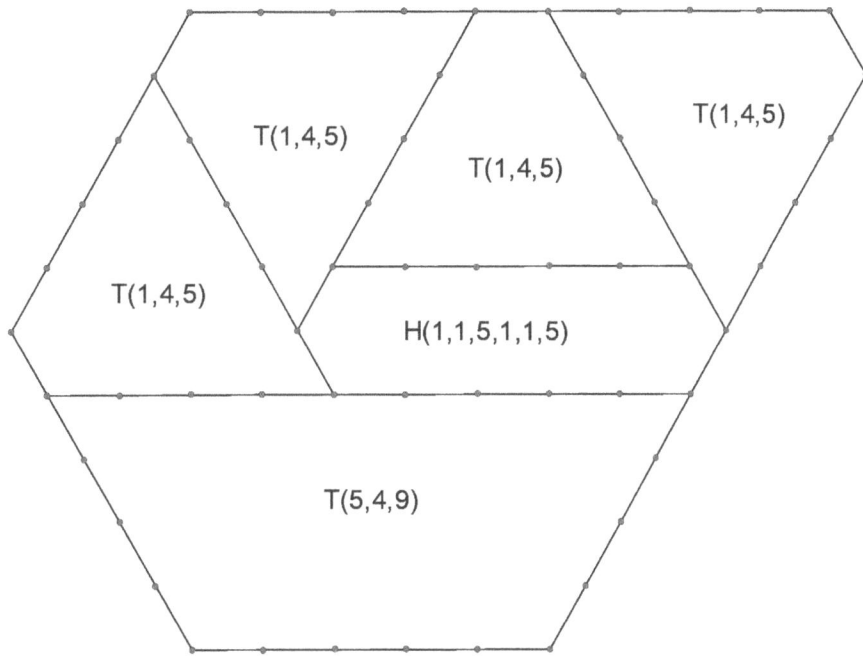

Figure 2.23

In general a trapezoid reduction takes the form $H(x,y, z, p, q, r) \rightarrow H(x-4, y+4, z-4, p, q, r)$ where x, z are larger than or equal to 5, and x, y, z are any three consecutive parameters. Suppose all 6 parameters are congruent to 3 mod 4 and H = H(23, 3, 15, 19, 7, 11), then a trapezoid reduction may proceed as follows

$(\underline{23}, \underline{3}, \underline{15}, 19, 7, 11) \rightarrow (11, \underline{15}, \underline{3}, \underline{19}, 7, 11) \rightarrow (11, 3, 15, \underline{7}, \underline{7},\underline{11}) \rightarrow$
$(\underline{11}, \underline{3}, \underline{15}, 3, 11, 7) \rightarrow (3, 11, \underline{7}, \underline{3}, \underline{11}, 7) \rightarrow (3, \underline{11}, \underline{3}, \underline{7}, 7,7) \rightarrow (3, 7, \underline{7}, \underline{3}, \underline{7}, 7) \rightarrow$
$(3, \underline{7}, \underline{3}, \underline{7}, 3, 7) \rightarrow (3, 3, 7, 3, 3, 7).$

The last member of this sequence cannot be further simplified by trapezoid reduction, but it is a nilpotent semiregular hexagon by Theorem 2.12. As a consequence, the original hexagon H, is also nilpotent.

In case at least one of the parameters of H is congruent to 1 mod 4, and at least one parameter is congruent to 3 mod 4, then trapezoid reduction will yield a nonnilpotent hexagon as in the following example with H = H(3,13,1,5,11,3).

$(3, \underline{13}, \underline{1}, \underline{5}, 11, 3) \rightarrow (3, 9, \underline{5}, \underline{1}, \underline{11}, 3) \rightarrow (3, \underline{9}, \underline{1}, \underline{5}, 7, 3) \rightarrow (3, 5, \underline{5}, \underline{1}, \underline{7}, 3) \rightarrow$
$(3, \underline{5}, \underline{1}, \underline{5}, 3, 3) \rightarrow (3, 1, 5, 1, 3, 3).$

For the last member of this sequence an application of the linear pattern can be used to show that it represents a nonnilpotent hexagon. As a consequence the original hexagon H is also nonnilpotent.

2.8 Hexagons with sides of mixed parity

If the hexagon H = H(p,q,r,s,t,u) has all even sides, then it has a tripod dissection, and thus is nilpotent. So we will suppose in this Section that H has at least one even and one odd length side. A second form of reduction is the following. If any two opposite sides of H, say p, s, are both larger than 4, then the reduction, which we call a *compression*,

$$(p, q, r, s, t, u) \rightarrow (p - 4, q, r, s-4, t, u)$$

can be performed. Every hexagon with a side of length greater than 7 is amenable to our two reductions. The only hexagons with sides of length 6 or 7 of this type which are not amenable to further reduction are

H(7, 1, 4, 4, 4, 1), H(6, 2, 3, 4, 4, 1), H(6, 1, 4, 3, 4, 1), H(6, 1, 3, 4, 3, 1).

Among these, only H(6, 2, 3, 4, 4, 1) is nilpotent as illustrated with a prime factorization in Figure 2.24. The further reduction of trapezoid (ABCD) in Figure 2.24 should be allowed:

$$(6, 2, 3, 4, 4, 1) = (\underline{4}, \underline{1}, \underline{6}, 2, 3, 4) \rightarrow (0, 5, 2, 2, 3, 4)$$

Here, (0, 5, 2, 2, 3, 4) represents a nilpotent pentagon. Also among our reductions the following *parallelogram reduction* could be included:

$$(6, 2, 3, 4, 4, 1) \rightarrow (6, 2, 1, 6, 2, 1).$$

In this case (DEFG) in Figure 2.24 is the severed parallelogram.

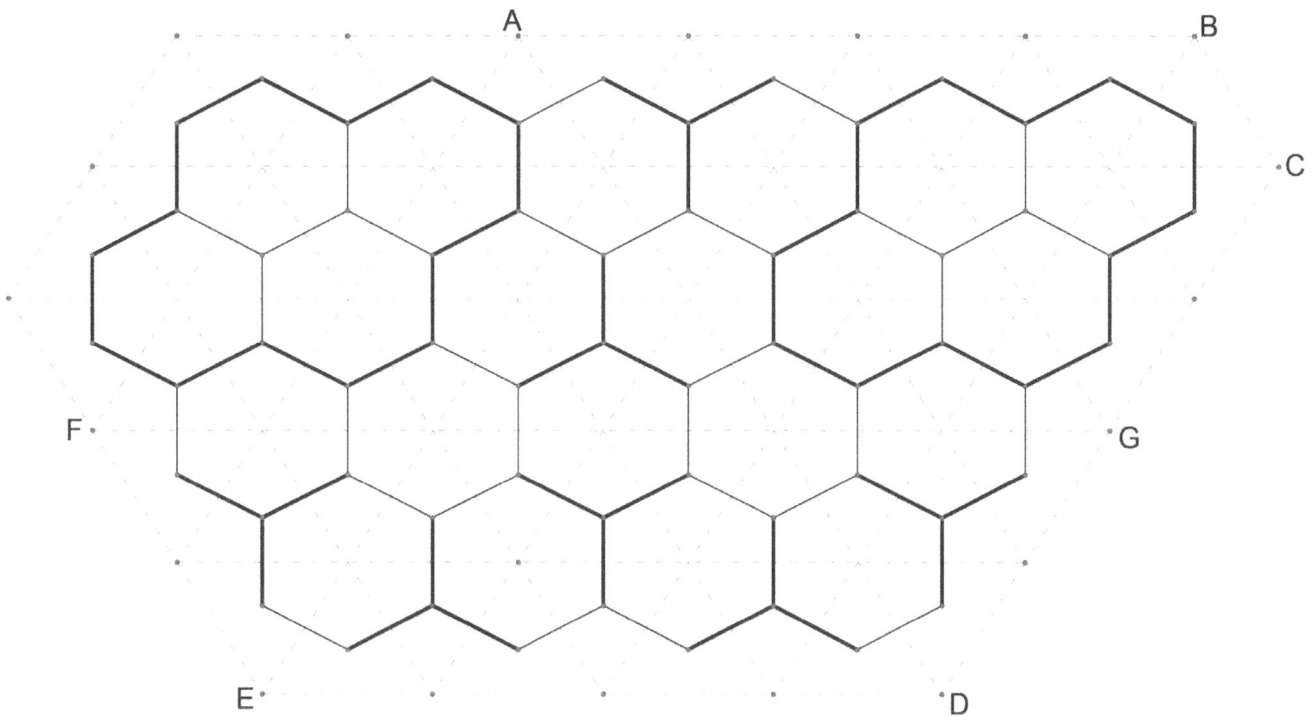

Figure 2.24

In considering the many possibilities for hexagons with all sides less than 6, the following considerations are useful. Since our attention is confined to hexagons H with at least one odd side and one even side, we can suppose that in H = H(p,q,r,s,t,u) p is odd and q is even. Equations (1),(2) of Section 2.7 yield the following three possibilities for parity sequences.

(I) (odd, even, even, even, odd, odd)
(II) (odd, even, odd, even, odd, even)
(III) (odd, even, even, odd, even, even).

Case (I) is illustrated in Figure 2.25. By appending the nilpotent triangles Δ_s, Δ_q to sides s, q respectively the trapezoid T(r+s+q, s+t, u) is created. Since s+t is odd, in this Case, this trapezoid is nonnilpotent by Theorem 2.5. The nonnilpotence is established using the linear pattern, and since this pattern assigns a value of 0 to the appended triangles, we may conclude that H is itself nonnilpotent.

47

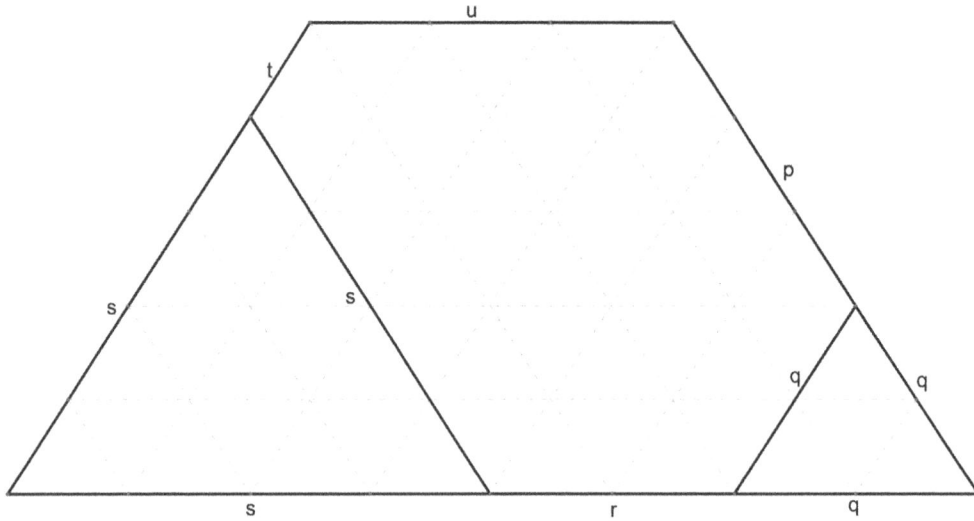

Figure 2.25

A similar argument can be used to show that hexagons of type (II) are nonnilpotent.

Case (III) is illustrated in Figure 2.26 in which we have appended the nilpotent triangles Δ_q, Δ_u to the even sides q, u respectively. The result is the trapezoid T(p+q+u, r+q, s) which is nilpotent by Theorem 2.5 iff the lateral side r+q is a multiple of 4. In Figure 2.26 this construction is illustrated for the hexagons H(3,2,6,1,4, 4) and H(1,2,6,1,2,6). A prime factorization of each hexagon is indicated.

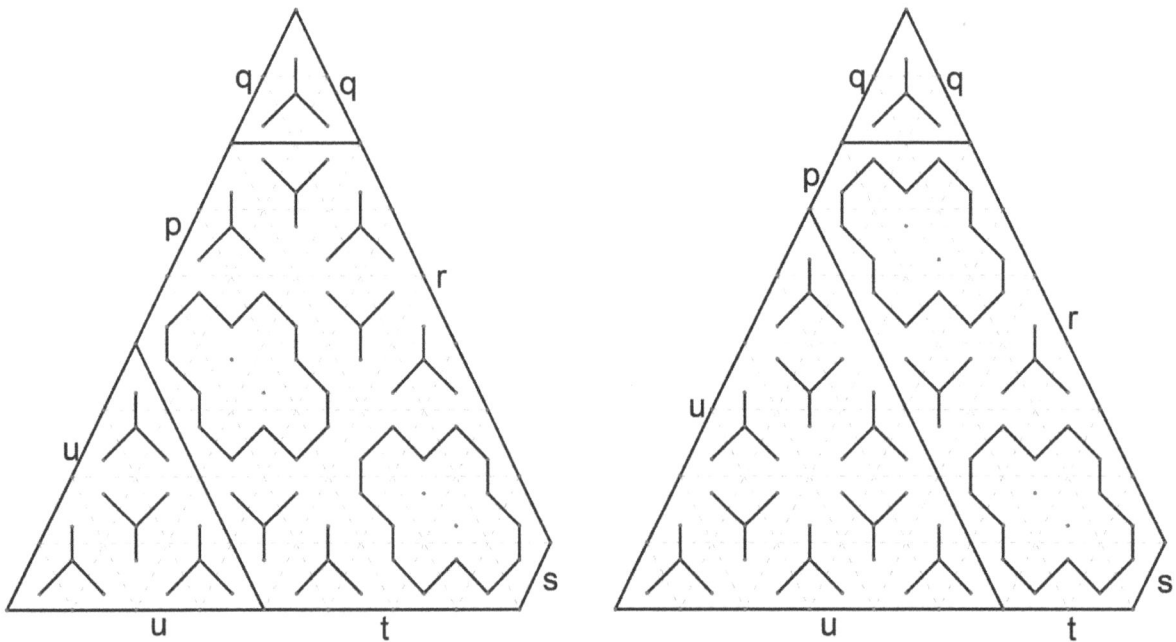

Figure 2.26

48

2.9 Character

By the *character* of a tiling of the plane we mean to refer to characteristics of a locally nilpotent assignment made to the entire plane. If for any such assignment, two basic regions must have exactly the same boundary values in the same relative positions, then we say the two regions are *equivalent*. Thus equivalence constitutes an important characteristic of the plane under consideration.

A *fundamental region* in a tiling $\prod$ of the plane, is a collection of equivalent basic regions which cover the entire plane $\prod$ via translations. The plane $\prod_2$ has a fundamental region consisting of an H(1,4,4,1,4,4) hexagon, as bound in bold in Figure 2.27. In this Figure we have indicated the chains of point couples which result in the recreation of the boundary values of hexagons A_0, A_2, $A_{5,}$ A_9 in the hexagons on the lower left side of the fundamental region. The same sort of duplication of values occurs on all sides of the fundamental region, and indeed the fundamental region itself along with its values covers the entire $\prod_2$ plane via translations.

In Figure 2.27, the values of the six points which lie on a *zig-zag* beginning with p and ending with -u generate the entire fundamental region, and thus generate the locally nilpotent assignment for the entire $\prod_2$ plane. We say that $\prod_2$ has *dimension 6*. This observation allows the determination of all patterns involving only values in the set {0,1,-1}. So far we have encountered the linear pattern, (Figure 2.4), and the M_6 and tripod pattern illustrated in Figure 2.8. These patterns provide a simple method for proving nonnilpotence. A third pattern called the *wedge and rhomb pattern* in [4] is illustrated in Figure 2.28. This pattern can be used to show that the hexagons H(1,3,1,3,1,3), H(1,1,3,1,1,3), H(3,3,1,3,3,1), H(5,1,3,3,3,1) are nonnilpotent.

The only remaining nontrivial pattern over {0,1,-1} is illustrated in Figure 2.29; it involves only the values {1,-1}. (There is, of course a *trivial pattern* which assigns 0 to every vertex.)

In determining the dimension of $\prod_2$ it is necessary to assert the associated basic nilpotent configurations, namely H_1 and the tripod. If we associate with $\prod_2$ just the hexagon H_1 as the only basic nilpotent configuration, then it is easily determined that the dimension is infinite. (The dimension of the plane Π_1 in Chapter 1, is also infinite. In Chapter 3 we will consider nilpotent properties of the plane Π_2 with the hexagon H_1 as the only basic nilpotent configuration.)

Another important characteristic of a tiling Π of the plane has to do with the actual procedure involved in the creation of a locally nilpotent assignment. In Π_1 and Π_2 the locally nilpotent assignment can be *freely* made. This is to say that as long as all forced values are assigned, then a new value, (adjacent to an assigned point), can be freely chosen with the ultimate result of a complete locally nilpotent assignment.

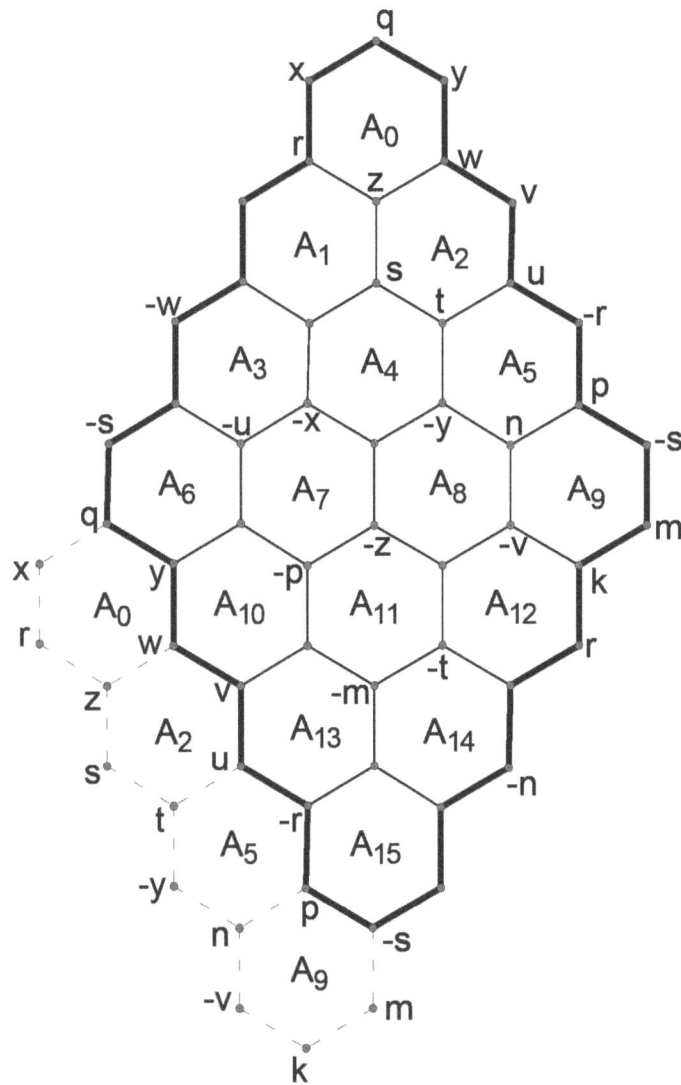

Figure 2.27

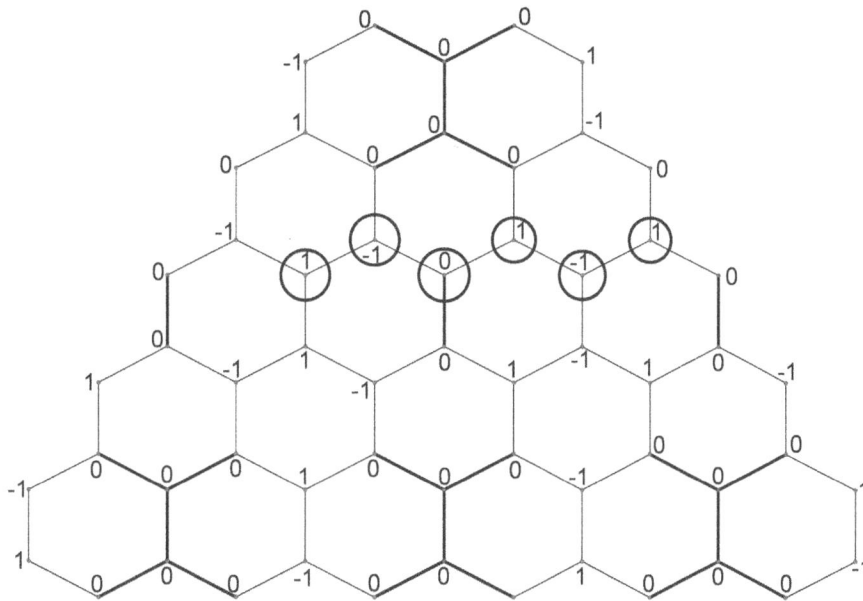

Figure 2.28

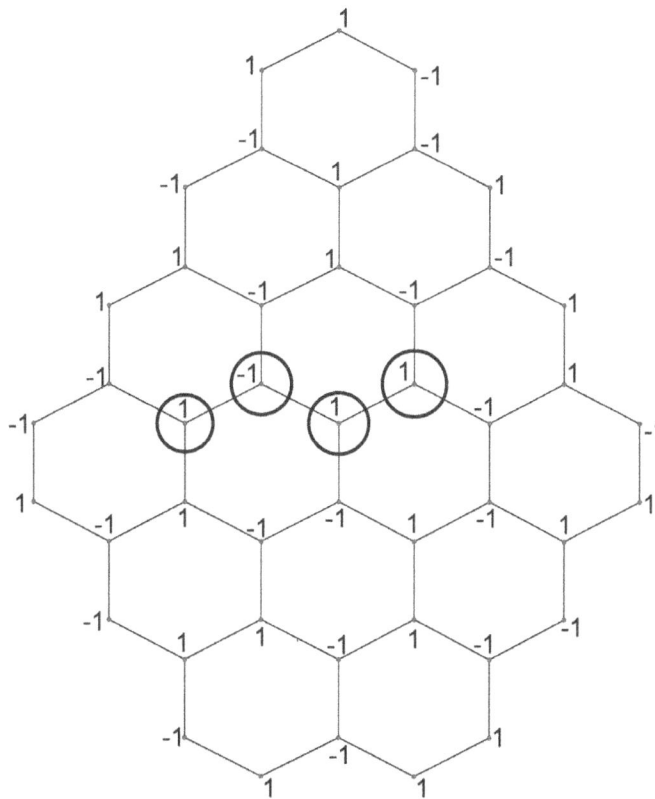

Figure 2.29

In Figure 2.30 a section of a tiling of the plane is illustrated. In this example a free locally nilpotent assignment *cannot* be made. In Figure 2.30(I) variables have been assigned representing a possible locally nilpotent assignment.

51

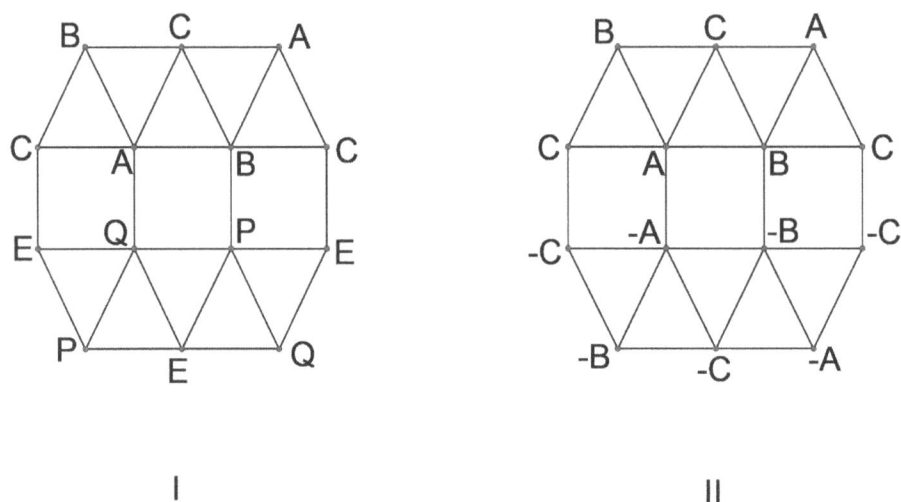

Figure 2.30

The initial labeling in Figure 2.30(I) reflects the fact that points at opposite ends of the long diagonal of a rhomb which, consisting of two adjacent equilateral triangles, must have the same value. Local nilpotence then requires that the following system of equations be solved.

(1) $A + B + C = 0$
(2) $A + C + E + Q = 0$
(3) $A + B + P + Q = 0$
(4) $B + C + P + E = 0$
(5) $P + E + Q = 0.$

These equations yield $P = -B$; $E = -C$; $Q = -A$, as illustrated in Figure 2.30(II). The structure in this Figure is indeed nilpotent, but nilpotent structures for planes with no free assignment have not, in our experience, provided for interesting games.

2.10 Extensions

In this Section we briefly consider two methods of extending a nilpotent polygon to a larger nilpotent structure. In Figure 2.31 two nilpotent extensions of H(1,2,2,1,2,2) are illustrated.

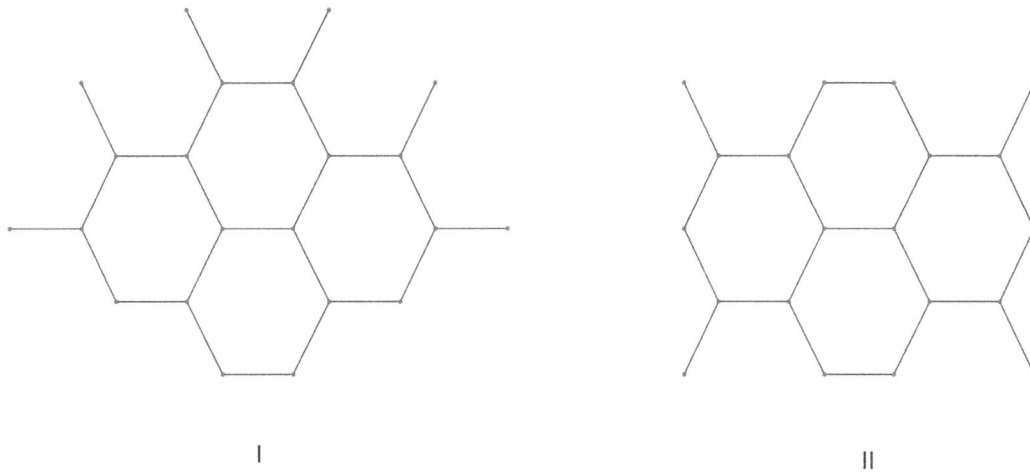

Figure 2.31

In Figure 2.32 a *point extension* of H_2 is illustrated. Several different point extensions of H_2 yielding nilpotent constructions are actually possible:

(1) All 12 extending points.
(2) Any 6 consecutive extending points, e.g., {A,B,C,D,E,F,G}.
(3) Any 6 alternating points, e.g., {A,C,E,G,I,K}.

In Figure 2.33 a second instructive point extension is illustrated.

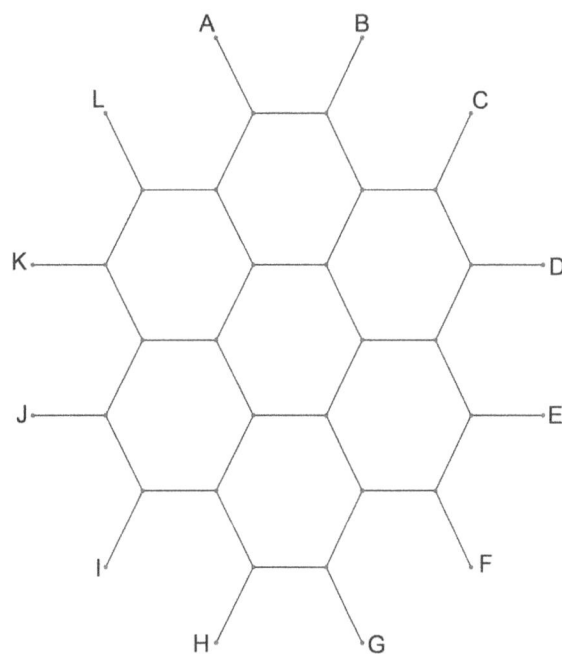

Figure 2.32

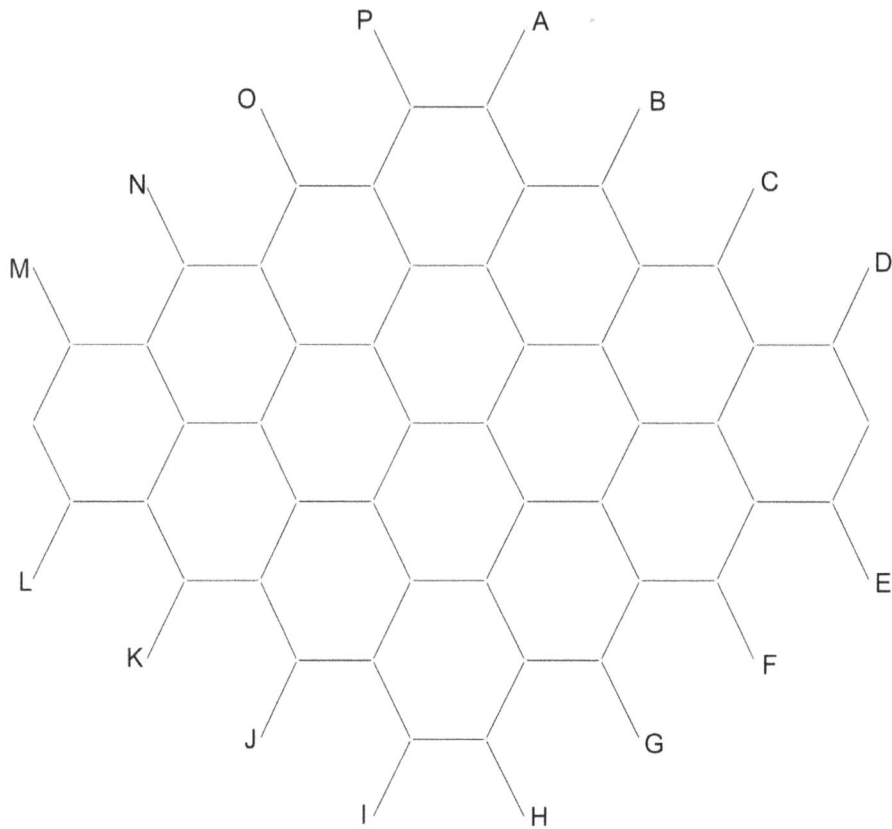

Figure 2.33

Several possible point extensions are possible in this example also:

(1) Each of the quadruples {A,B,C,D}, {E,F,G,H}, {I,J,K,L}, {M,N,O,P} have sum 0, so any or all can be added, yielding nilpotent extensions.
(2) The point couples {N,C}, {K,F} can be added separately, or together.
(3) The six points {M,O,P,A,B,D} as well as {E,G,H,I,J,L} have sum 0, and also constitute nilpotent extensions.

Another type of extension involves pure polyhexes. The nilpotent structure in Figure 2.34 we call a *tripod* and denote by TR(a,b,c) where the parameters represent the lengths of the three linear polyhexes which meet at the central hexagon. So Figure 2.34 illustrates the tripod TR(5,5,5).

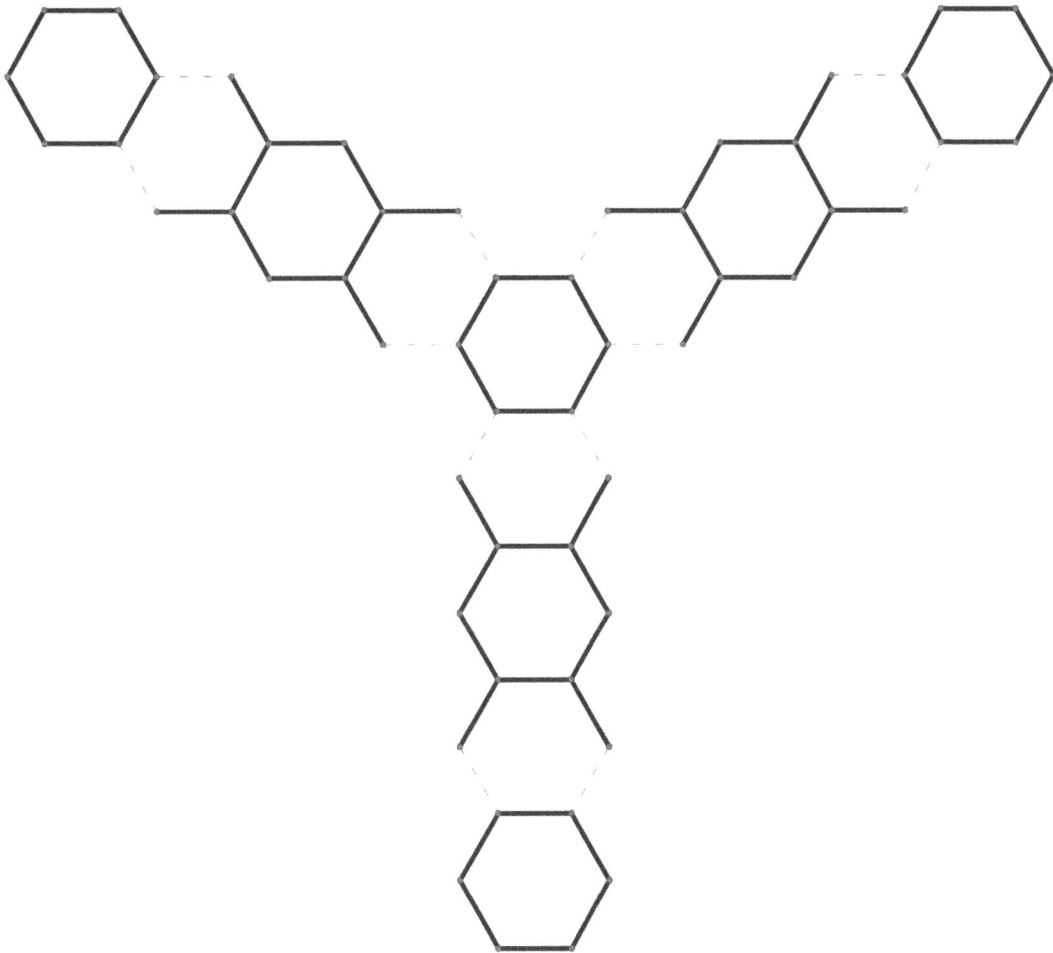

Figure 2.34

The arms of the tripod can be extended independently using linear polyhexes of length 4. Thus every tripod of the form TR(1+4s, 1+4t, 1+4u) can be constructed. In fact, nilpotent tripods of the form TR(a+4s, a+4t, a+4u) can be constructed for a ∈ {0,1,2,3}. There is no limit to the nilpotent polyhexes which can be obtained by extending tripods; we just provide one example in Figure 2.35. The uncovered points A,B,C,D,E,F are easily seen to sum to 0.

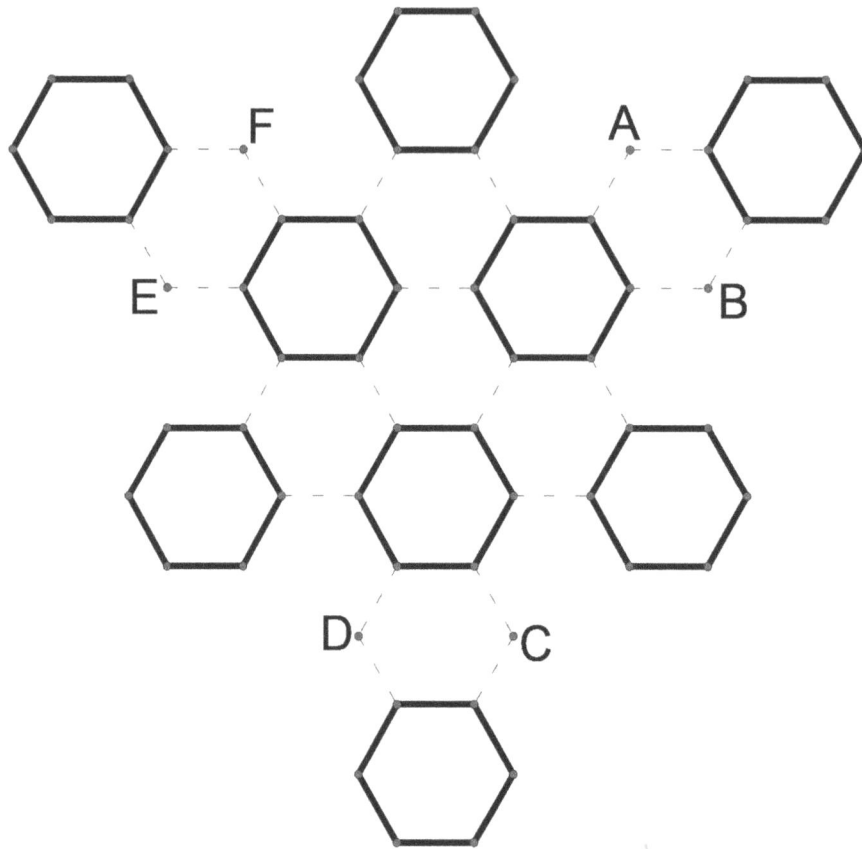

Figure 2.35

It is an interesting problem to determine if a given nilpotent configuration is prime or composite. As an example, consider the following factorization of one of the extensions of the nilpotent polyhex illustrated in Figure 2.33. The factorization in Figure 2.36 clearly shows that the polyhex along with the point extensions A, B is nilpotent, (B is equal to -A by means of a point couple chain).

Now cut the graph as indicated by the curve (c,d). This yields two factors: a nilpotent M_2 and the remainder of the graph which we will just denote by W. For our cut to yield a prime factorization, it must be that W is nilpotent when considered in isolation. The problem is that the point couple chain linking points A, B has been broken. In order to show that W is nilpotent it suffices to show that, even with this chain broken, we still have B = -A. To this end consider the labeled polyhex in Figure 2.37. In this Figure we have already labeled the various point couples in accordance with Theorem 2.1.

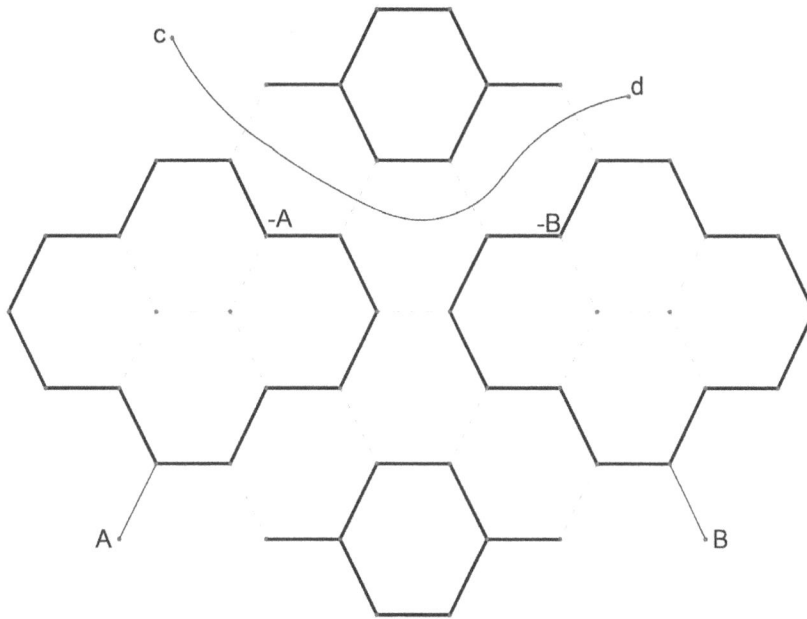

Figure 2.36

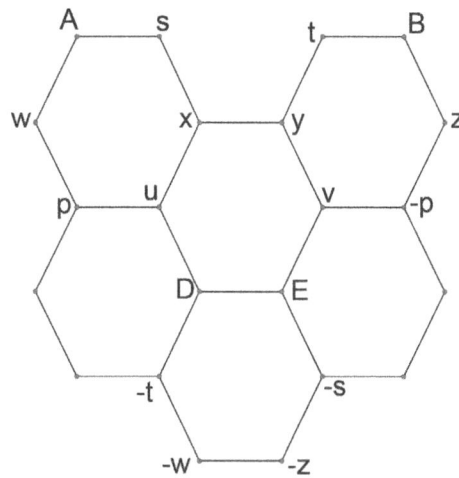

Figure 2.37

Considering the hexagons containing A, B we obtain

(1) $A + s + x + u + p + w = 0$,

(2) $B + t + y + v + z - p = 0$.

Now replacing $(x + u + p)$ in (1) by -D, and replacing $(t + y + v)$ in (2) by -x we obtain

(3) $A + s + w - D = 0$,

(4) $B - x + z - p = 0$.

57

Solving (3), (4) for A, B respectively we obtain

(5) A + B = (D − s - w) + (x + p − z).

Now using Lemma 2.3 to replace -w − z in (5) by -x − y, we obtain

(6) A + B = D − s + p − y.

Finally, using tripods to replace D − s by -E -v, and then -E − v by y − p, we obtain

(7)   A + B = -E − v + p − y = y − p + p − y = 0.

It now follows that, despite the point couple chain being broken in W, we still have B = -A, and thus W is indeed nilpotent. No further prime factorization of W is possible, and thus W is prime.

2.11 The game

The Π_2 *game* will be played on the vertices of the graph illustrated in Figure 2.38. This graph has 102 points. As usual scoring configurations will consist of theoretically nilpotent subgraphs. 100 playing discs are used in 10 groups, with values {0,1, …, 9} in each group. A score is obtained when the discs occupying a nilpotent configuration sum to a multiple of 5. In the game "nilpotent" means congruent to 0 mod 5, and the score is whatever multiple of 5 is achieved,

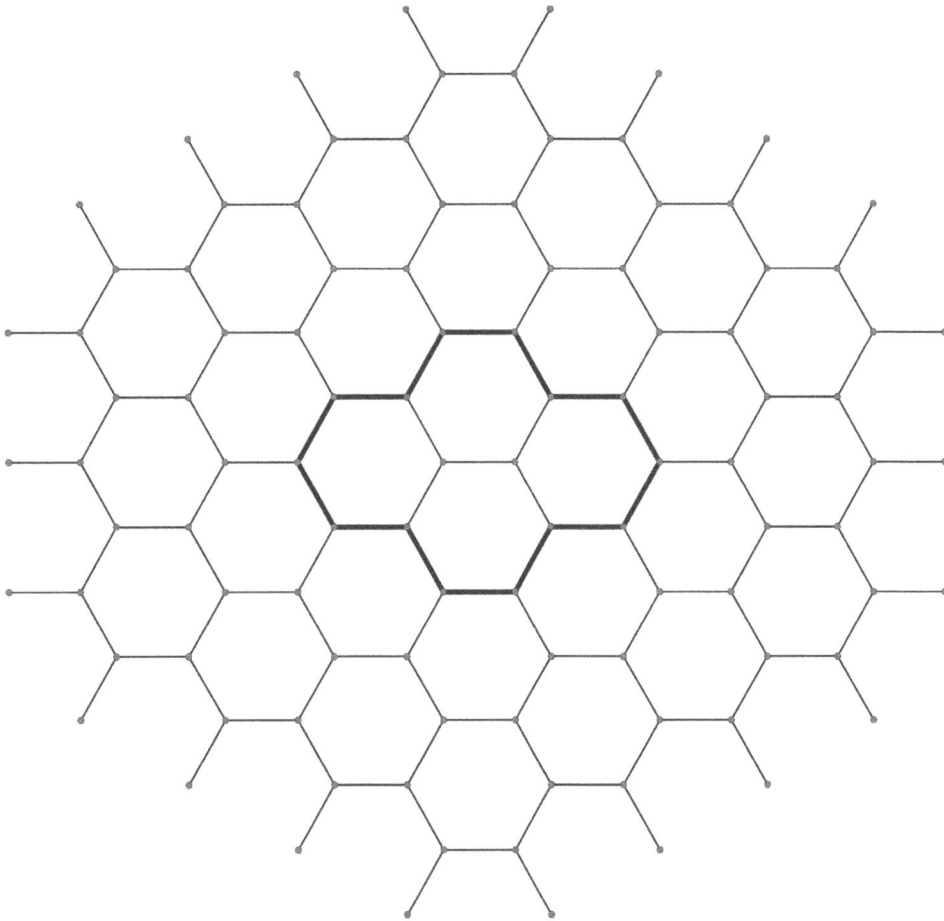

Figure 2.38

There are so many nilpotent structures in $\prod_2$ that, for practical reasons, it is important to restrict to only a few. We think all the following should be included $\{\Delta_2,$ (the tripod), $H_1, M_2, M_4, M_6, H(1,2,2,1,2,2), H_2, \Delta_4, L_5, H(2,2,3,2,2,3)\}$. Larger nilpotent triangles or hexagons are not likely to occur in the play of the game. Nilpotent parallelograms and trapezoids do occur but we do not think they should be included as scoring configurations.

Also we advise against including nilpotent point extensions other than the mites and perhaps G_6, (Figure 2.21), and the two $H(1,2,2,1,2,2)$ extensions in Figure 2.31. The tripod $TR(2,2,2)$ will occur frequently, but almost always in conjunction with an $H(1,2,2,1,2,2)$ or an H_2. We advise not to include this tripod as a scoring configuration.

We have not discussed conditionally nilpotent $\prod_2$ polygons in the previous Sections. Some of these may also be included as scoring configurations. In Figure 2.39 some of the smaller conditionally nilpotent hexagons are illustrated.

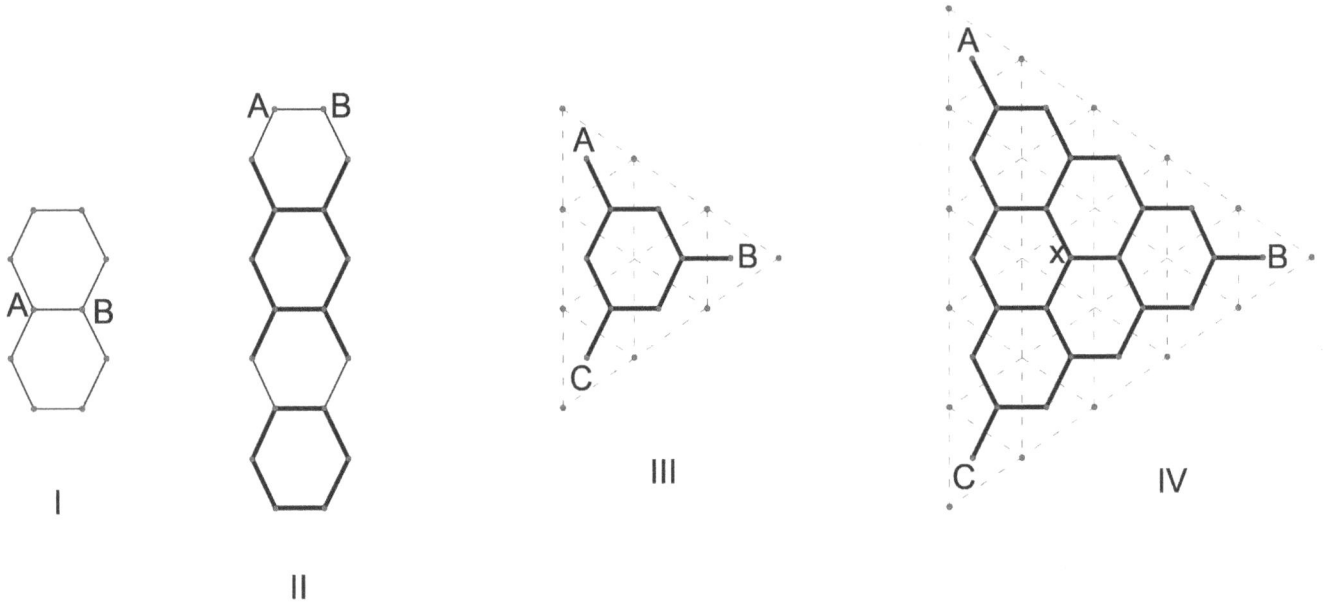

Figure 2.39

In Figure 2.39, the two linear polyhexes, L_2^*, L_4^* are 0-valued iff A+B = 0. In Figure 2.39(III) the triangle Δ_3^* has a zero value iff A+B+C = 0. In 2.39(IV) the triangle Δ_5^* has a zero value iff the central point x is 0. Note that we indicate conditionally nilpotent polygons using an asterisk.

As in the rules for the $\prod_1$, the two players each select 10 discs from the boneyard. One further disc is selected from the boneyard and is placed in one of the two central points of the playing board. The selected discs constitute the *hand* of each player. A players *turn* consists of placing one or more discs on the playing board always keeping a connected *central configuration*.

There are two types of placements: (i) a scoring position, or (ii) a nonscoring position. A scoring position involves a placement of a disc which completes either a tripod or an H_1 hexagon. In such positions *only a scoring disc can be placed*. A nonscoring position involves placing a disc adjacent to a point in the central configuration which does not complete a tripod or an H_1 and which produces no additional score.

In case a nonscoring play creates an adjacent dead point, a *blocking play* results. The Π_2 game can be played in two versions: blocking plays allowed or blocking plays not allowed. The first version adds to strategic possibilities but due to the creation of dead points, the larger scoring configurations are less likely to occur. In the second version some strategic possibilities are eliminated but larger scoring configurations are more likely to occur.

In a single turn a player may make as many scoring placements as possible, and then complete his/her turn by placing a nonscoring tile. In the event that all 10 discs have been placed in scoring positions, the players turn also ends. *Only when the player's turn ends,* may he/she replenish the hand to a full count of 10 discs by drawing from the boneyard.

Players alternate turns until no further legal placements are possible. If one player has no legal moves or has exhausted his/her hand, then the player *passes,* and the opponent has an opportunity to play. The player with the highest score is the winner.

As in Π_1 *dead points* occur when a point occurs at the junction of two tripods, or a tripod and a hexagon and which require two different values in order to create a multiple of 5. Such positions should be marked, (we use a button). No disc can be placed on a dead point and no scoring configuration can contain a dead point. Players can create dead points in order to reduce scoring possibilities for the opponent.

It is also possible that the placement of a disc simultaneously satisfies the scoring requirements of two tripods, or a tripod and a hexagon. In this case score is earned for each completed configuration. Quite often two scoring configurations are completed by the placement of a single disc. *Unless one configuration is completely contained in the other,* the full scores for both configurations are tabulated. This *double score* occurs frequently when an M_6 and a Δ_4 are the scoring configurations involved. This situation is illustrated in Figure 2.40.

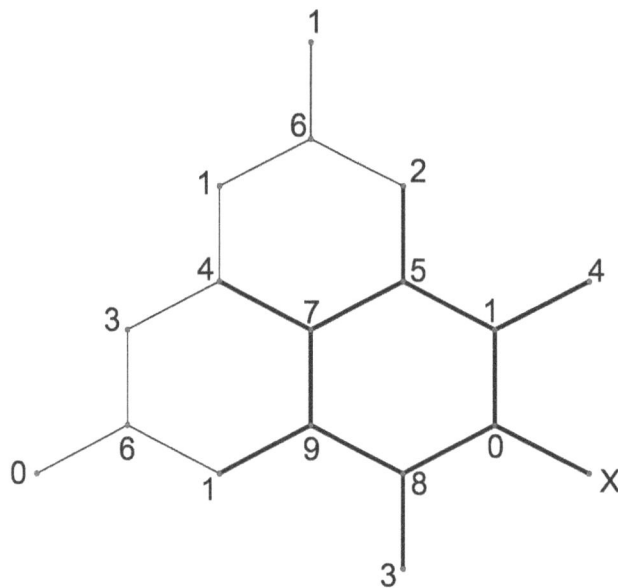

Figure 2.40

If a disc with value 6 is placed at position X in Figure 2.39, then both a Δ_4 with score of 60 points, and an M_6, indicated in bold, with a score of 50 points are completed. Since the M_6 contains two points, (the point extensions with values 3, 4), which are not included in Δ_4, the total score for the play at point X is 110 points.

As in the Π_1 game there is a boundary, indicated in bold in Figure 2.38, in which all points must be covered by discs or dead points before any discs can be placed outside the boundary. In case a player cannot make a legal move inside the boundary, then he/she must pass. If two consecutive passes occur then the first player to pass may begin play outside the boundary.

A nonscoring disc which creates an adjacent dead point is called a *blocking play*. The Π_2 game can be played with or without blocking.

All of our games can be constructed without too much difficulty using mat board, wooden discs and number stickers. The wooden discs and number stickers can be found in a craft store. Appropriate numbers are placed on ea ch disc. To construct a hexagon lattice for the Π_1 or Π_2 playing boards, a compass, a protractor, a straight edge and a ruler is all that is required. 1.5 inches is a good length for the edge of a hexagon.

CHAPTER 3

Nilpotent $\prod_3$ subgraphs

3.1 Introduction

Recognizing the importance of stipulating the basic nilpotent configurations along with the accompanying plane, we introduce the following notation. For the plane $\prod_2$ of Chapter 2, we distinguish it by setting $\prod_2 = [\prod_2; H_1, \Delta_2]$. In this Chapter we consider $\prod_3 = [\prod_2; H_1]$; thus we use the same hexagon lattice $\prod_2$ as in the previous Chapter, but now only H_1 will serve as basic nilpotent configuration. The elimination of Δ_2, (the tripod), as a basic configuration produces a radical change in the structure of nilpotent configurations.

The first observation is that no connected subgraph of $\prod_2$ with a point extension can be nilpotent. This is so, since without the tripod, any value can be assigned to the extending point. The nilpotent configurations, then, will consist purely of polyhexes.

3.2 Prime $\prod_3$ polyhexes

In Figure 3.1 we consider the possibility that a prime polyhex consists of two adjacent hexagons. At left in this Figure, a locally nilpotent assignment using {0,1,-1} is illustrated. This assignment clearly extends to the entire $\prod_3$ plane by just assigning 0's to the entire remainder of the plane. The pair of adjacent hexagons A, B have a value of 1 under this assignment, showing that, taken together, they do not form a nilpotent polyhex. In order to obtain a possibly nilpotent structure, the hexagon C must be adjoined.

But then, a similar locally nilpotent assignment featuring a single hexagon island labeled with 1's and -1's, surrounded by an infinite ocean of 0 values is constructed, as at right in Figure 3.1. With this assignment the polyhex consisting of hexagons A,B,C is seen to be nonnilpotent, and in order to possibly obtain a nilpotent structure, hexagon D must be adjoined. This argument can be repeated ad infinitum without ever obtaining a nilpotent polyhex. The conclusion is stated in the following result.

Theorem 3.1 *The single hexagon H_1 is the only prime Π_3 polyhex.*

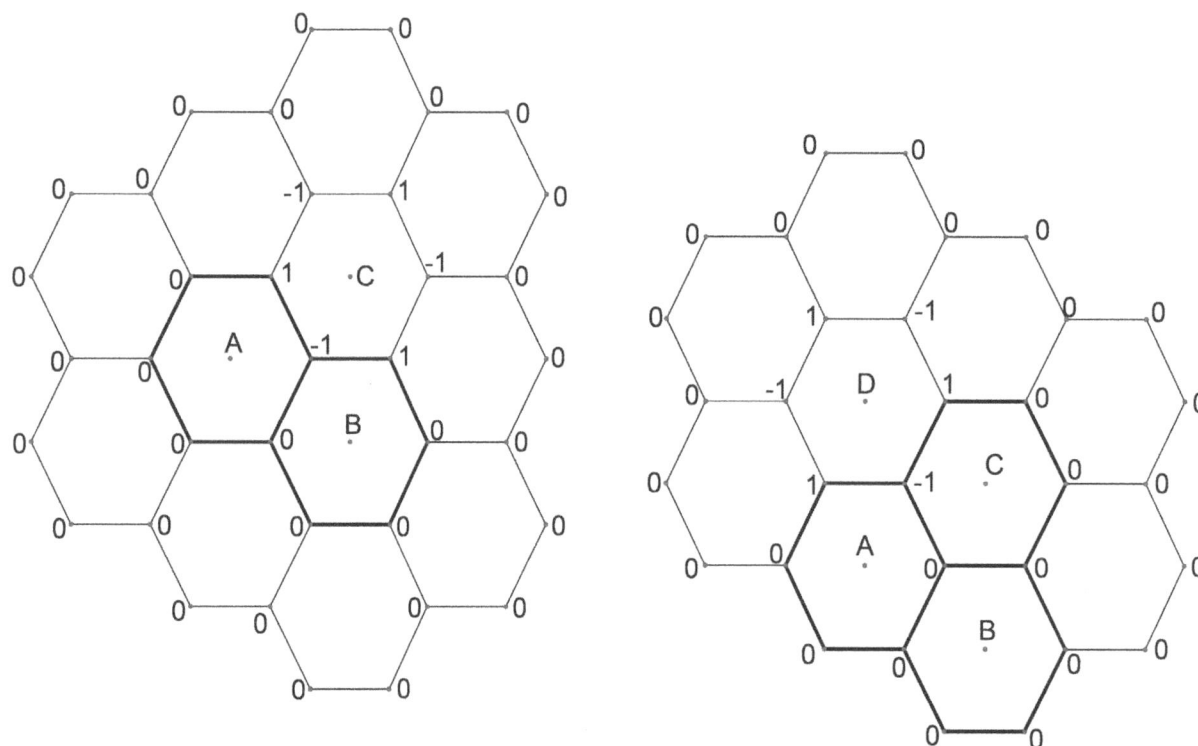

Figure 3.1

3.3 Composite Π_3 polyhexes

As a consequence of Theorem 3.1, composite Π_3 polyhexes must consist of a number of isolated H_1 hexagons. We will now consider the possibility that two such hexagons occur in a stack of three as in Figure 3.2. In this Figure we suppose that hexagons A, B are prime and hexagon D is intervening. The locally nilpotent assignment assigns a value of 1 to the stack {A,D,B} so this stack is nonnilpotent. The string of alternating

1's and -1's extending from left to right in Figure 3.2, can be extended without limit thus insuring that our assignment is universally locally nilpotent.

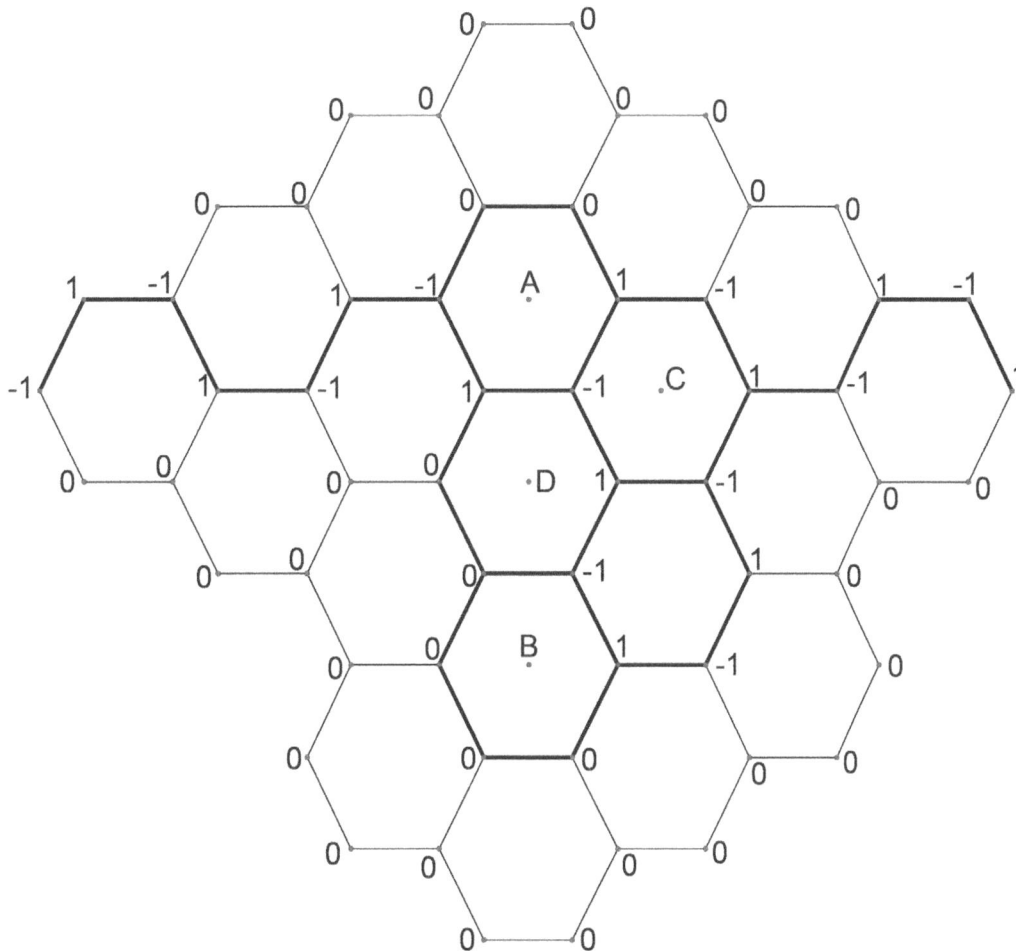

Figure 3.2

The conclusion is that any two of our prime H_1 hexagons cannot occur in a stack with just one intervening hexagon. As a consequence, in any composite $\prod_3$ polyhex, the prime constituents must be allocated as A, B in Figure 3.3. The locally nilpotent assignment in this Figure shows that the polyhex {A,B,E,F} is not nilpotent. Note that the locally nilpotent assignment extends without limit as a sequence of alternating 1's and -1's from the points x, y.

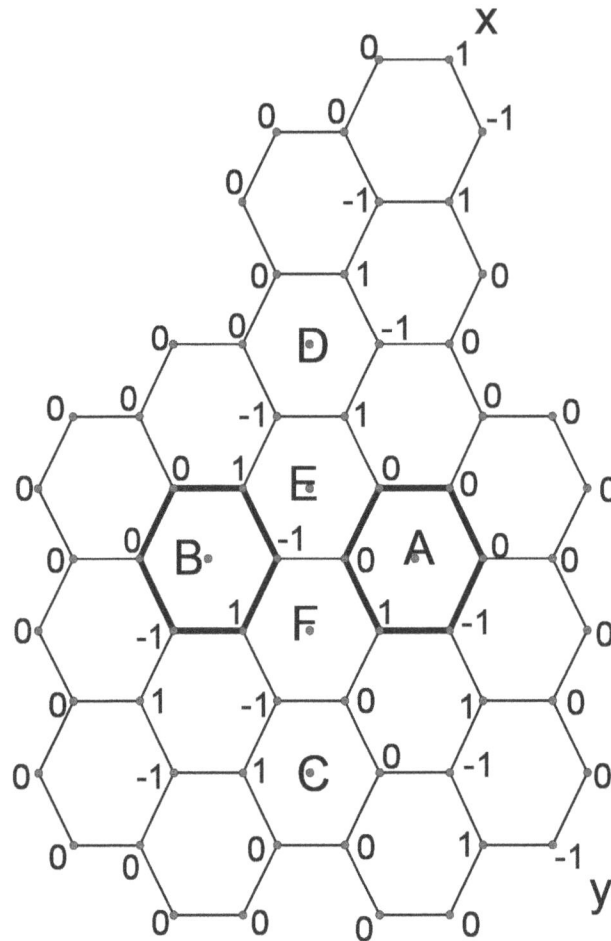

Figure 3.3

As a consequence either hexagon C or D must be included in order to obtain a composite polyhex. We have demonstrated the following result.

Theorem 3.2 *A composite $\prod_3$ polyhex is an extension of the tripod TR(2,2,2).*

In Figure 3.4 some examples of composite $\prod_3$ polyhexes are illustrated.

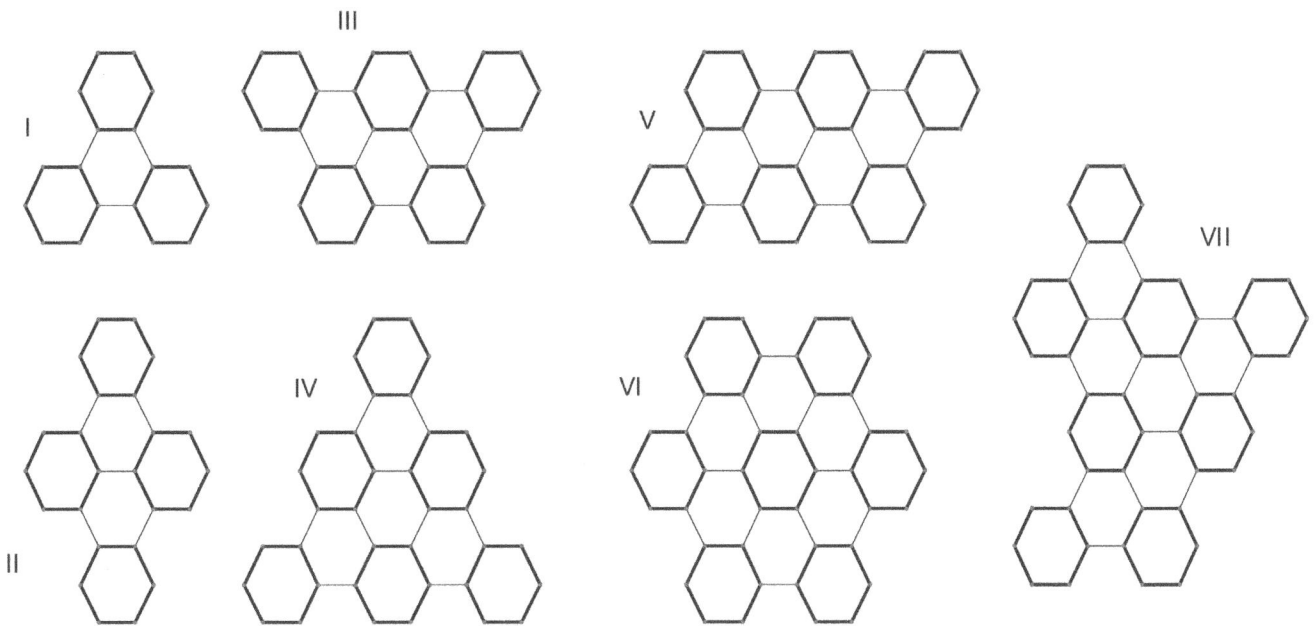

Figure 3.4

3.4 The Π_3 game

The Π_3 game will be played on the board illustrated in Figure 3.5. It has 80 points and is identical with the Π_2 board except for the lack of point extensions. Scoring configurations will be limited to H_1, the *tripod*, (Figure 3.4(I), the *quad*, (Figure 3.4(II), the *triangle,* (Figure 3.4(IV) and the *wheel*, (Figure 3.4(VI). Larger nilpotent configurations, like the one illustrated in Figure 3.4(VII) go against one of the main attractions of this game: its *simplicity*. Each hexagon is given an algebraic label for the purpose of ease in scoring.

There are 80 discs accompanying the playing board. They are numbered 0, ..., 9 in eight groups. These are placed face down in the boneyard and players select 10 discs for their hand from the boneyard. One more disc is taken from the boneyard and placed face up on one of the two center points. Players alternate turns and try to realize one of the scoring configurations. As usual a scoring play involves realizing one of the configurations with a sum which is a multiple of 5. The main difference from earlier games is that each player can play two discs at once.

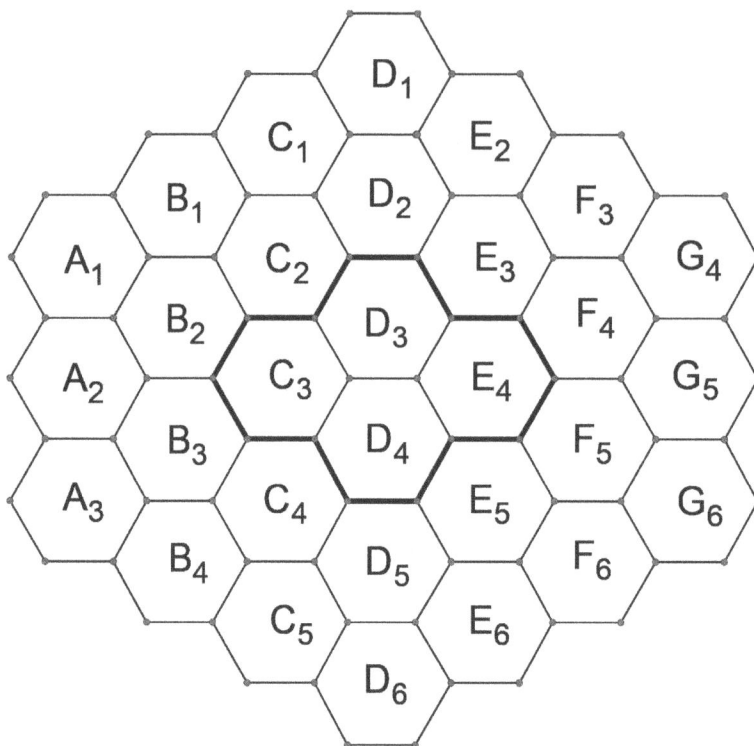

Figure 3.5

In a single turn a player continues to make scoring configurations by playing one or two discs as long as he/she chooses. Whether one or two discs are used, the player is entitled to play another pair of discs. The player terminates his/her turn by playing one disc in a nonscoring position, always adjacent to a played disc. It is not allowed to play a nonscoring disc or pair of discs at point(s) which complete a hexagon without achieving a multiple of 5.

In case a player has only one disc in hand after completing a series of scores, then that single disc is played in a nonscoring position and the players turn ends. The turn also ends in case a player has used all his/her 10 discs in scoring positions. Only at the end of the turn is a player allowed to replenish his/her hand to a full complement of 10 discs by selecting from the boneyard.

The first several played discs must be placed on points inside or on the boundary of the barrier indicated in Figure 3.5. This requirement stimulates the quick development of scoring positions.

Another distinguishing feature of this game is the almost complete lack of dead points which arise in the normal course of play. This allows for the abundant achievement of the large scoring positions, namely the triangle and the wheel.

Double scores, i.e. a play which completes more than one scoring configuration often occur in the natural course of play. The most common double score occurs when two quads are created simultaneously. If we follow the procedure for double scores given in earlier games, then the complete score for both configurations would be tabulated and summed. But for the sake of simplicity we adopt a different rule for $\prod_3$: if two scoring configurations are completed simultaneously, then we just tabulate the score for one of them. If two quads are completed, then the player should just take the one which yields the largest score. If a triangle and a wheel are completed simultaneously then the score determined by the wheel must be selected since it is the larger configuration.

To simplify the tabulation of scores, scorecards like the one illustrated in Figure 3.6 can be used.

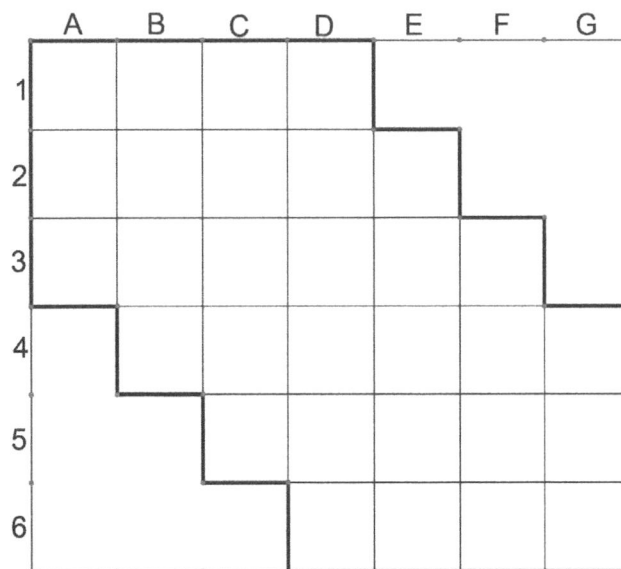

Figure 3.6

The scorecard has an entry, in accordance with the labels in Figure 3.5, for each of the hexagons on the playing board. Whenever a player completes a hexagon, the score earned is placed in the corresponding entry.

Two additional scorecards with the same structure are assigned to each player. Players just put a mark in their scorecard in the position(s) corresponding to hexagon(s) lying in any achieved scoring configuration. In this way only the original score of a hexagon is recorded during the course of the game.

When the game is finished then each players score is determined by multiplying the number of marks with the score for the corresponding hexagon, and then summing all such computations.

CHAPTER 4

Π_4 polygons

4.1 Introduction

A section of a tiling of the plane which we call Π_4 is illustrated in Figure 4.1. It has been provided with a locally nilpotent assignment generated by the adjacent values D, A, B, C, E on the top line. Π_4 thus has dimension 5. Also in Figure 4.1 the fundamental region is indicated using dashed lines. It contains 24 values and repeats via translations to cover the entire Π_4 plane.

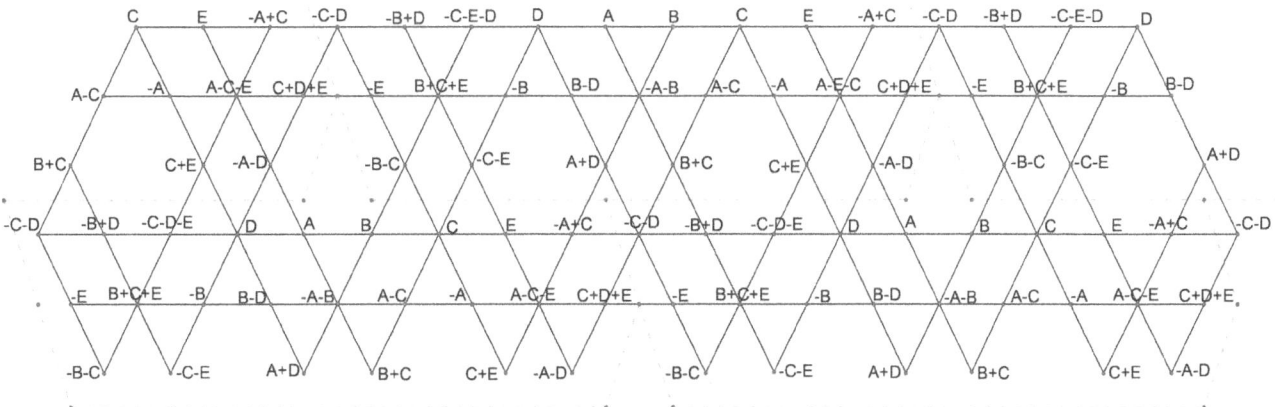

Figure 4.1

The basic nilpotent structures in the Π_4 plane consist of $\{\Delta_1, H_1, R_1\}$. Here, R_1 denotes a rhomb, (rhombus), with unit length sides and with interior angles of 60 and 120 degrees. The notation R_m will be used for a rhomb of the same type with sides of length m. As in the previous Chapters, we will study nilpotent Π_4 parallelograms, triangles, trapezoids and hexagons. Recall that nilpotent polygons fall into two categories: prime and composite. Only in some cases will we consider conditionally nilpotent Π_4 polygons.

4.2 Basic results

The algebraic point values in Figure 4.1 yield several quick observations. The relevant Π_4 polygons are listed in Figure 4.2.

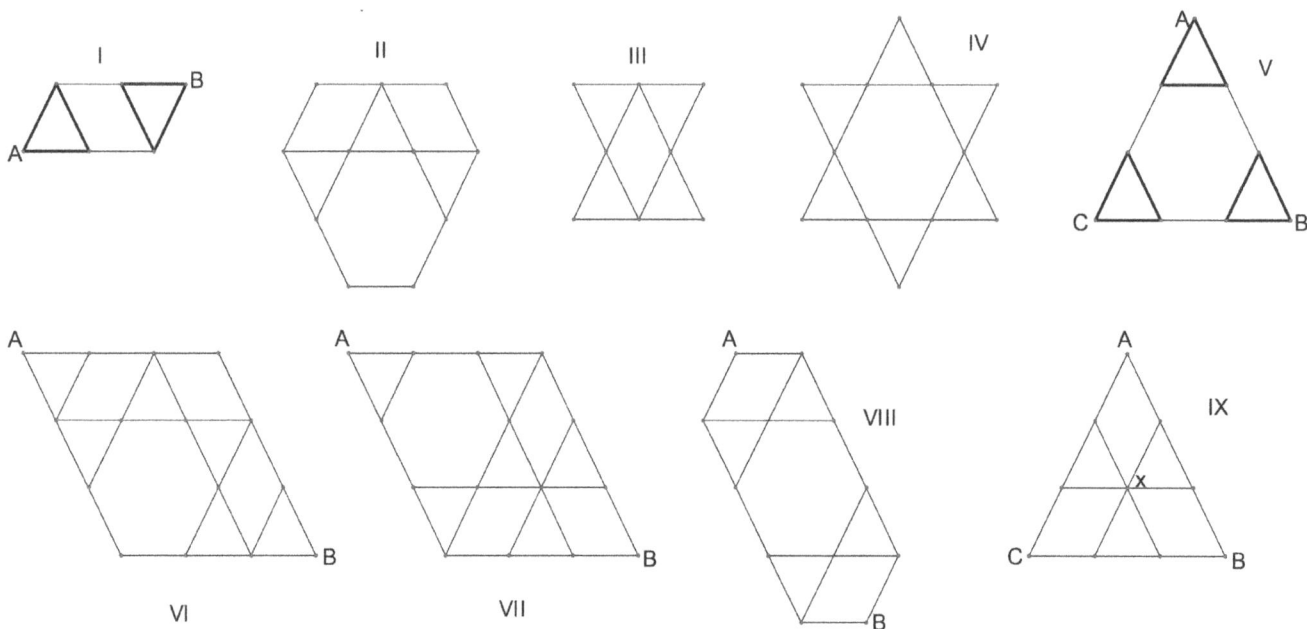

Figure 4.2

Theorem 4.1

(i)   *The P(1,2) parallelogram in Figure 4.2(I) is composite with A+B = 0.*

(ii)   *The H(1,2,1,2,1,2) hexagon in Figure 4.2(II) is prime.*

(iii) *The polygon in Figure 4.2(III) we call a wedge (W_I); it is prime.*

(iv) *Figure 4.2(IV) is a prime star, (J_I).*

(v)   *The Δ_3 triangle in Figure 4.2(V) is composite. We call it a Type I triangle to distinguish it from the Type II Δ_3 triangle in Figure 4.2(IX). The Type II triangle is conditionally nilpotent with the requirement that the central vertex x have a 0 value. In both types of triangles the sum of values of the three corner points A, B, C is 0 under any locally nilpotent assignment.*

(vi) The R_3 rhombs in Figures 4.2(VI, VII) are nonnilpotent, but in both cases the values of the endpoints A,B of the long diagonal are equal in any locally nilpotent assignment.

(vii) The H(1,1,3,1,1,3) hexagon in Figure 4.2(VIII) is nonnilpotent, but the opposite boundary points A, B sum to 0 in any locally nilpotent assignment.

From a study of Figure 4.1 it might be concluded that the points A, B in the H(1,1,4,1,1,4) hexagon in Figure 4.3(I) sum to 0 in any locally nilpotent assignment. But this is not the case, as the reader can easily determine. Instead, the larger context illustrated by the enclosing H(2,2,3,2,2,3) in Figure 4.3(II) is required in order to obtain A+B = 0. As usual we leave the proof of this fact, as well as most of the Theorem proofs as exercises.

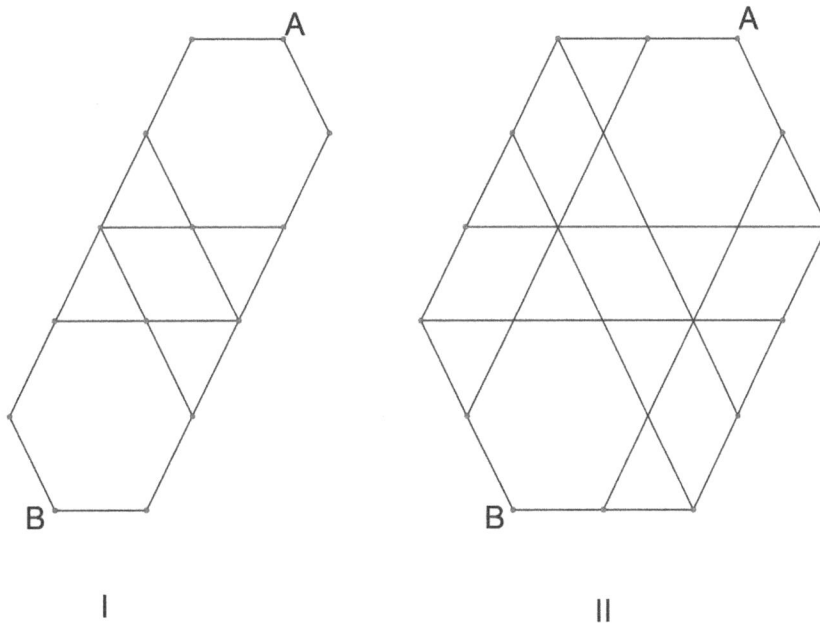

Figure 4.3

4.3 Introduction to nilpotent $\prod_4$ triangles

There is one Δ_2, two types of Δ_4's and two types of Δ_5 triangles in the $\prod_4$ plane, and none of these is nilpotent. There are two Δ_6 triangles illustrated in Figure 4.4.

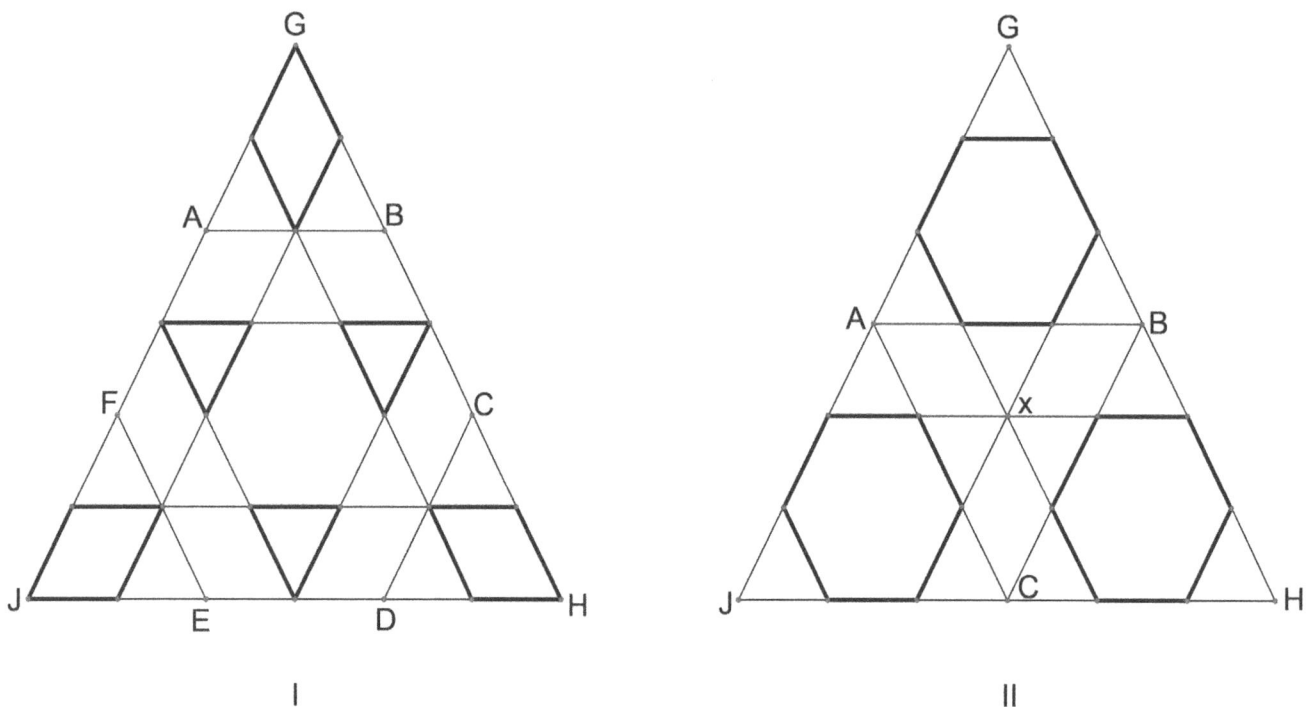

Figure 4.4

The Type (I) Δ_6 triangle in Figure 4.4(I) is prime, and the Type(II) Δ_6 triangle in Figure 4.4(II) is conditionally nilpotent with the requirement x = 0. The partial covers with basic tiles $\{\Delta_1, R_1, H_1\}$ indicate a procedure for proving these statements. In both Types, the sum of corner values G, H, J is 0 in any locally nilpotent assignment.

There is essentially one type of Δ_7 triangle in the $\prod_4$ plane. It is composite as indicated in Figure 4.5, where a factorization into a nilpotent Δ_3 triangle and a prime T(4,3,7) trapezoid is indicated. There is one Δ_8 triangle in the $\prod_4$ plane which we illustrate in Figure 4.6. For the purpose of analysis an algebraic locally nilpotent assignment has been included.

74

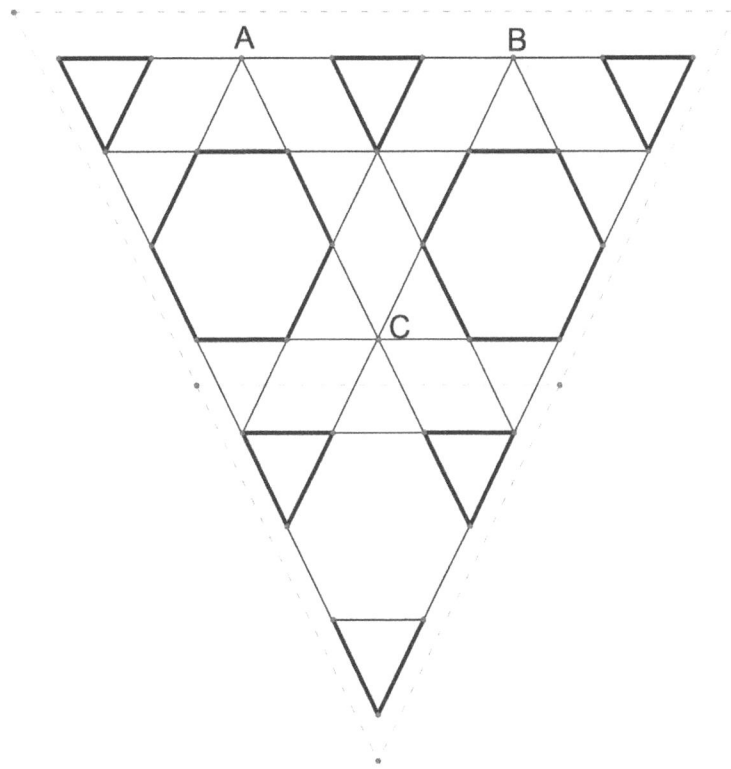

Figure 4.5

The point of interest in Figure 4.6 is that the sum of the values on the top row as well as the other two sides of Δ_8 is 0. By deleting the top row or the right side of Δ_8, a nilpotent Δ_7 results, and thus the Δ_8 is nilpotent.

The segment consisting of nine values with sum 0, we will call a *0-segment*. The nine point 0-segment occurs as the larger base of a T(4,4,8) trapezoid since at least such a trapezoid is required to generate the segment. If this trapezoid is extended by to a P(6,4,0,9,1,3) pentagon, then we find that the first and last point of the resulting segment of length 9, are identical. (The 0 in the notation for a pentagon indicates the location of the unique interior 60 degree angle.) This important observation we record as a Theorem.

Theorem 4.2 *The 9 points comprising large base of a T(4,4,8) trapezoid have sum 0 in any locally nilpotent assignment. If this trapezoid is extended to a P(6,4, 0, 9,1,3) pentagon, then the first and last point of the resulting boundary segment of length 9, have identical values.*

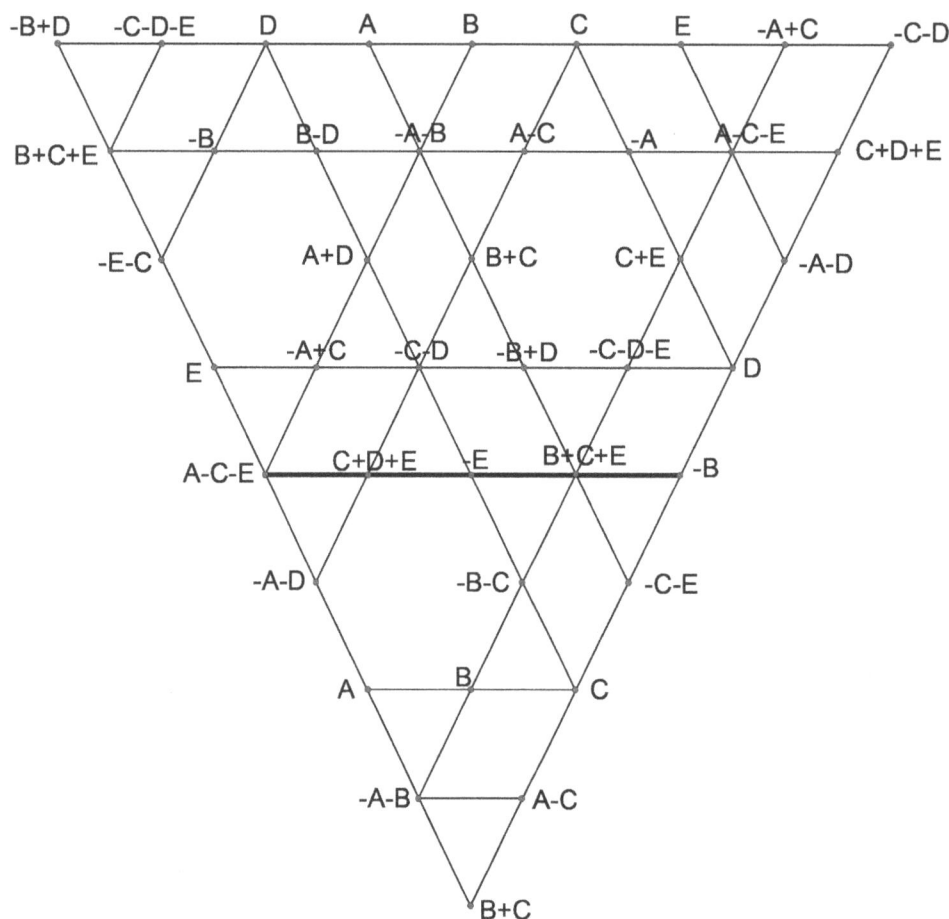

Figure 4.6

4.4 General categories of $\prod_4$ triangles

$\prod_4$ triangles can be put into three categories depending on the way in which the corner points are covered. The types of corners are illustrated in Figure 4.7. Category(I) triangles cover all three corners with Δ_3's like the one in Figure 4.7(I). Triangles in Category II cover at least one corner point with a Δ_2 triangle, as in Figures 4.7(II). In Category (III) triangles, all three corners are covered with R_1 rhombs via Δ_2 triangles as in Figure 4.7(III).

We will first study the Category(I) triangles. Category(I) triangles have the form Δ_{3+9t}.

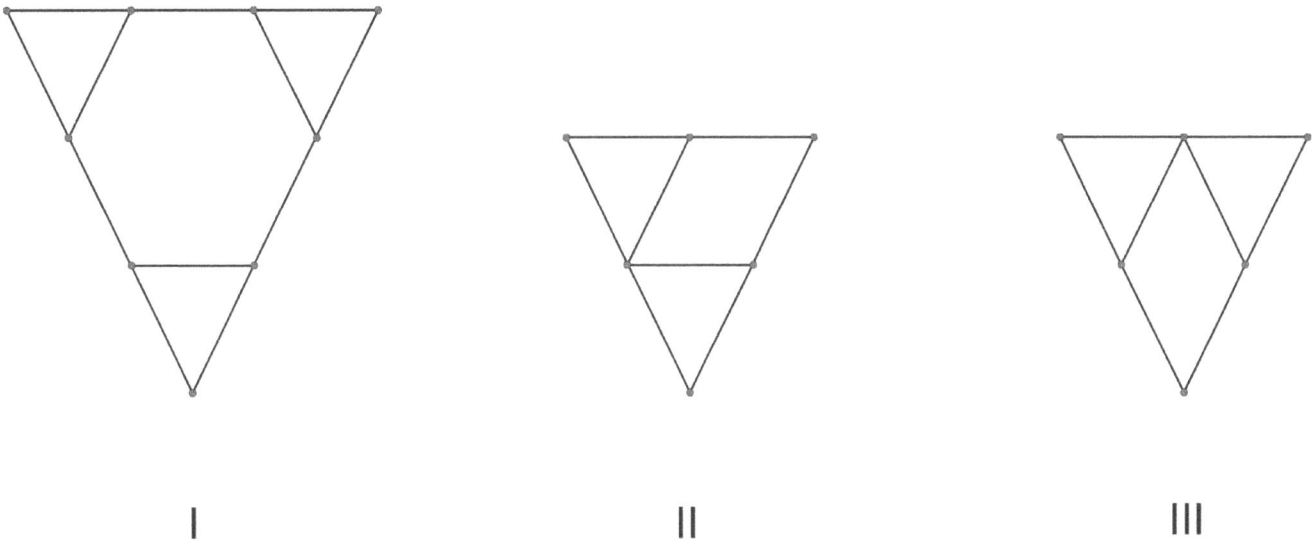

Figure 4.7

The Δ_3, as in Figure 4.7(I), is of course, nilpotent, so our first consideration is to prove the nilpotence of Δ_{12}. In Figure 4.8 a factorization of Δ_{12} is given which features an H_4 hexagon and three Δ_3's in the corners. In the H_4 hexagon we have $A_1 + A_2 + A_3 = 0$, and $B_1 + B_2 + B_3 = 0$ as in Figure 4.4. Otherwise H_4 is factorized using six H_1's and a J_1 star, and thus H_4 is nilpotent, and in fact, prime.

The factorization of larger triangles of the form Δ_{3+9t} continues in the pattern indicated by Figure 4.9. In this Figure the triangles represent Δ_3 and the hexagons represent H_4. Note that if each H_4 is *collapsed* to H_1 and each Δ_3 is collapsed to Δ_1, then we obtain nilpotent $\prod_1$ triangles. In fact, any nilpotent $\prod_1$ polygon can be *expanded* by reversing this process, to obtain a corresponding nilpotent $\prod_4$ polygon.

To determine the nilpotent character of Category(II) triangles, consider the locally nilpotent assignment given in Figure 4.10 for such a triangle. In this Figure a dissection into two Δ_7 triangles, an R_8 rhomb, and two 0-segments is indicated. The R_8 consists of 9 rows of which 6 are 0-segments and 3 *nonsegment rows* also with sum 0. The two Δ_7 triangles are equivalent and are nilpotent as in Figure 4.5.

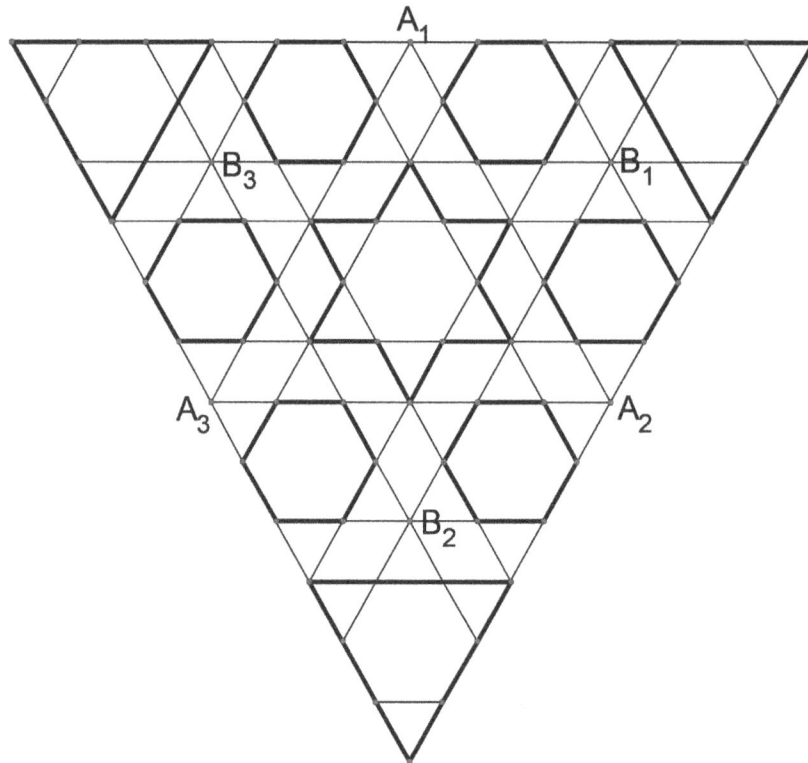

Figure 4.8

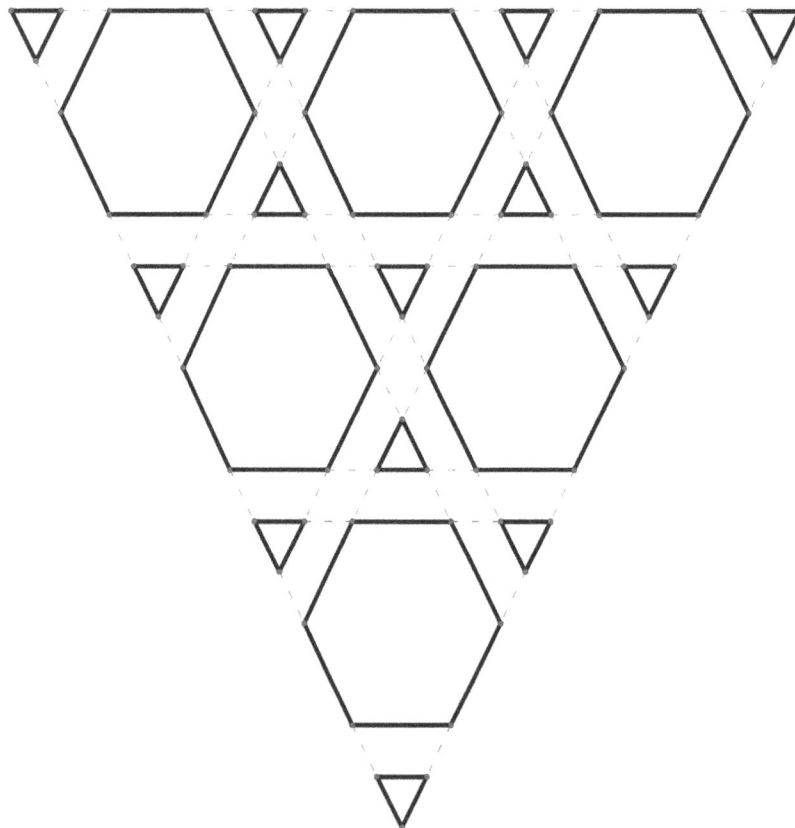

Figure 4.9

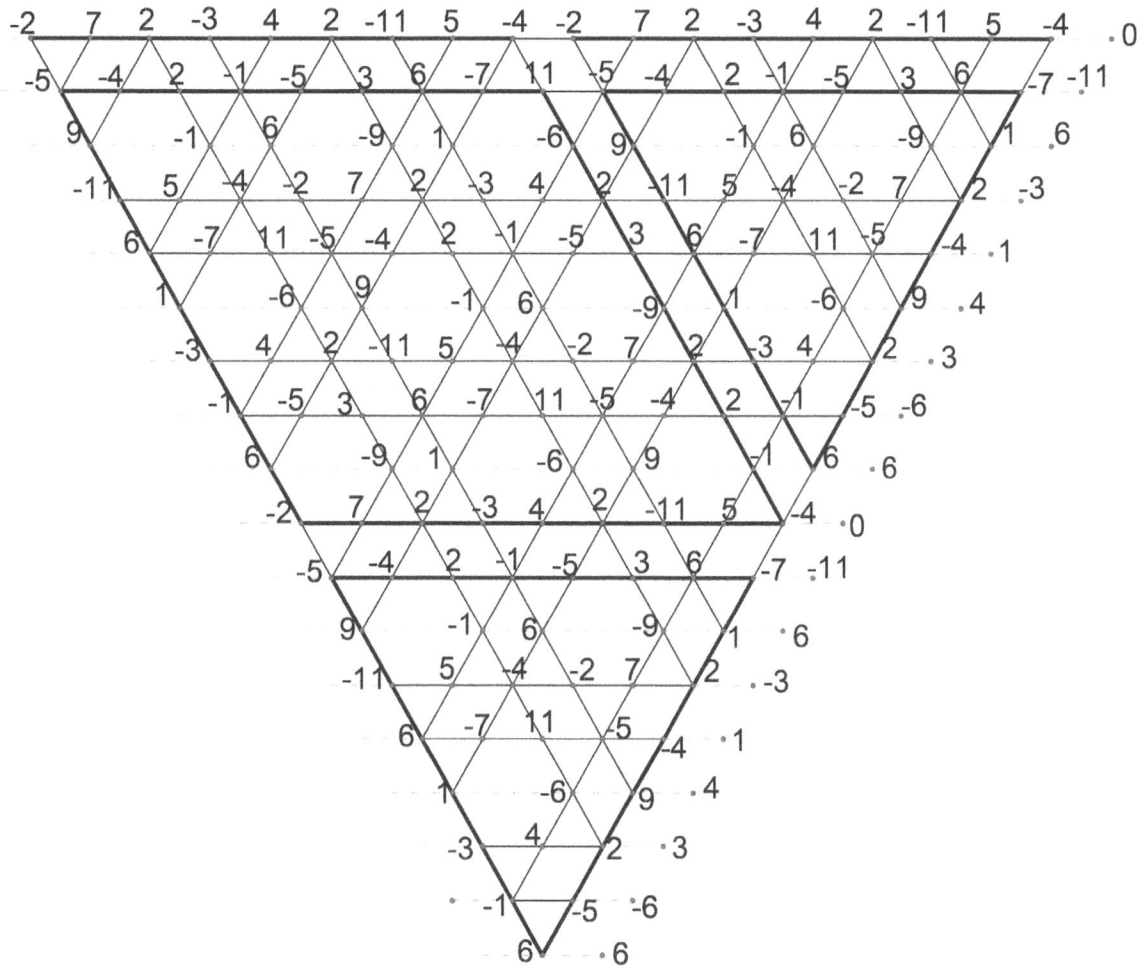

Figure 4.10

The R_8 rhomb is easily seen to be nilpotent, and thus the T(8,8,16) trapezoid in Figure 4.10 is composite.

As a consequence of these observations, the row sums, indicated for each row at right in Figure 4.10, repeat in segments of length 9. We have seen that Δ_8 is nilpotent, and we also have immediately that Δ_{16} and Δ_{17} are nilpotent. Also from the repeating nature of the row sums, we can conclude that Category(II) triangles of the form Δ_{2+9t}, Δ_{4+9t}, Δ_{5+9t} are nonnilpotent.

This leaves only the Category(II) triangles of the form Δ_{1+9t} to consider. In Figure 4.11 the nilpotence of Δ_{10} is illustrated. As a further consequence of the repeating structure of Category(II) triangles we also obtain that all Category(II) triangles of the form Δ_{1+9t} are nilpotent. Among these only Δ_1 is prime.

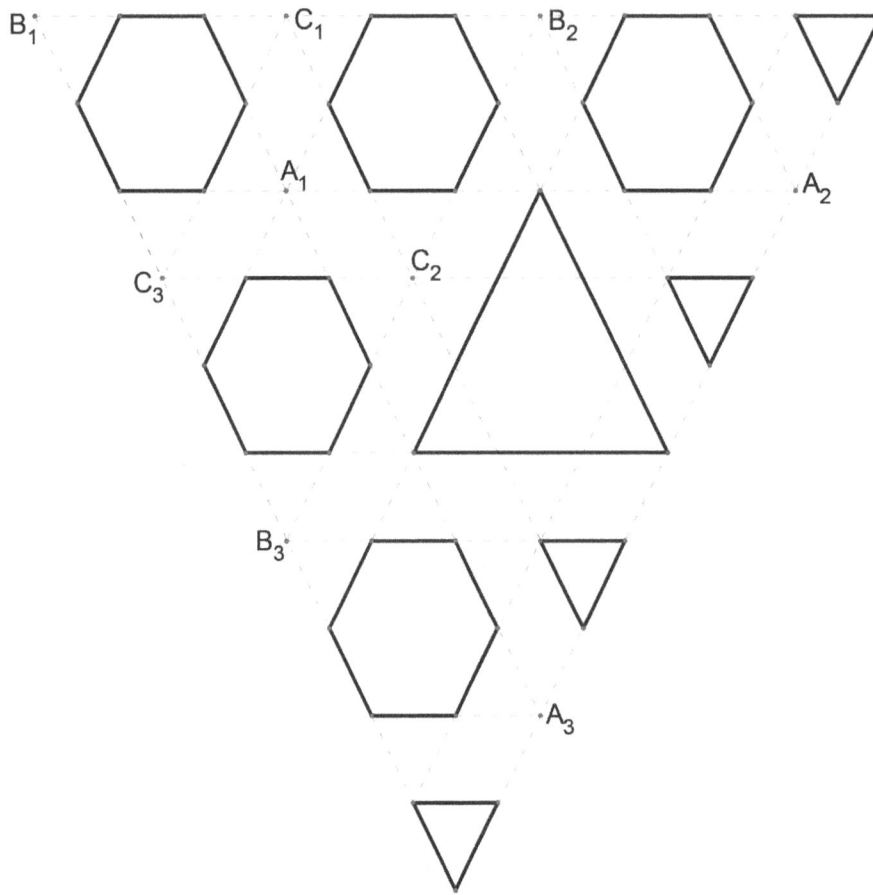

Figure 4.11

Triangles in Category(III) all have the form Δ_{6+9t} and all are nilpotent as we will now argue. Triangle Δ_6 illustrated in Figure 4.4(I) is prime. Δ_{15} is composite: it can be factored as a nilpotent Δ_8 and the nilpotent $T(9,6,15)$ trapezoid in Figure 4.12.

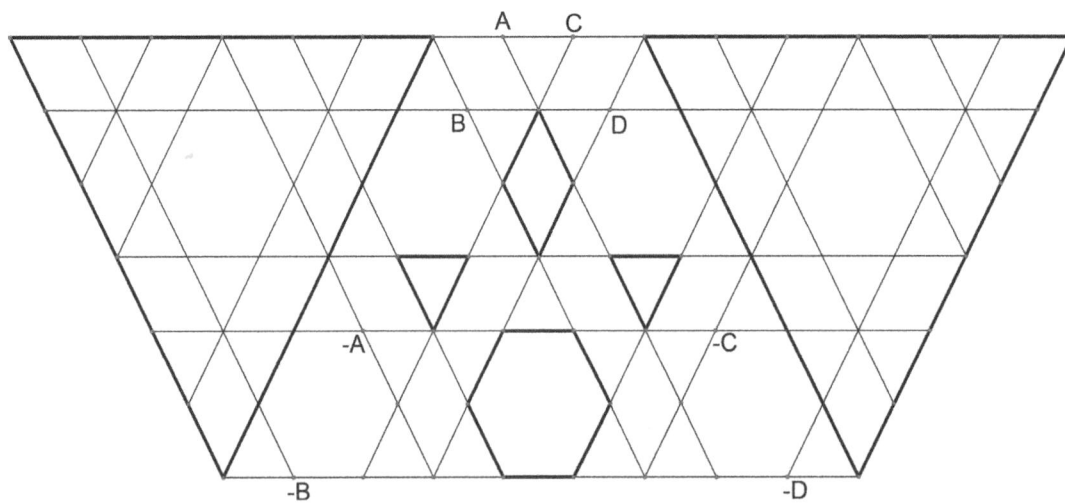

Figure 4.12

In general, we find that the nilpotent Category(II) triangle Δ_{8+9t} extends to the Category(III) triangle Δ_{15+9t}. This extension is accomplished via the factorization $\{\Delta_{8+9t},\ T(9(t+1),\ 6,\ 6+9(t+1))\}$. The T(18,6,24) trapezoid involved in the extension $\Delta_{17} \rightarrow \Delta_{24}$ is illustrated in Figure 4.13.

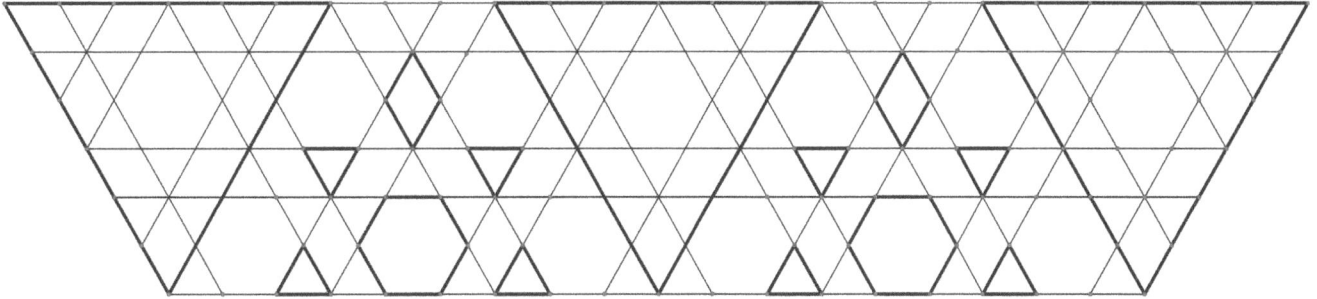

Figure 4.13

The first section of the trapezoid in Figure 4.13 is the nilpotent T(9,6,15) trapezoid of Figure 4.12. It is then extended by means of a section of T(9,6,15). This extension can be repeated ad infinitum to obtain any T(9t,6,6+9t) trapezoid.

The results of this Section are summarized in the following Theorem.

Theorem 4.3 *Category(I) triangles have the form Δ_{3+9t} and all are composite. In Category(II) only triangles of the form Δ_{1+9t}, Δ_{7+9t}, Δ_{8+9t} are nilpotent. Among these only Δ_1, Δ_7, Δ_8 are prime. Category(III) triangles have the form Δ_{6+9t}, and all are nilpotent. Among these only Δ_6 is prime.*

4.5 $\prod_4$ trapezoids with lateral sides 1, ..., 6

4.5.1 $\prod_4$ trapezoids with lateral side 1 or 2

We have already seen the importance of trapezoids in the determination of nilpotent $\prod_4$ triangles. In this Section we will investigate trapezoids methodically in terms of the length of the lateral sides. There are two types of $\prod_4$ trapezoids with lateral side 1 as illustrated in Figure 4.14.

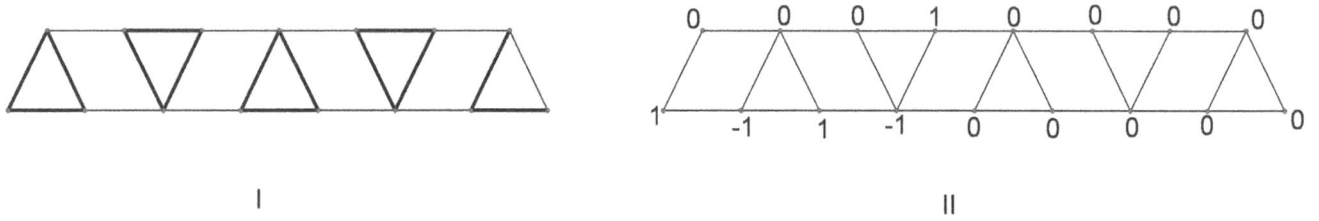

Figure 4.14

Trapezoids of the form T(3t, 1, 3t+1) as in Figure 4.14(I) are all nilpotent and composite for t ≥1 since they have a Δ_1 factorization. Trapezoids T(1,1,2), and T(m, 1, m+1), m ≥4 are all seen to be nonnilpotent by means of the {0,1,-1} locally nilpotent assignment as in Figure 4.14(II). Note that the tail of this trapezoid is all 0's, and since the initial segment has sum 1, our contention for m ≥4 immediately follows. The reader can easily find a different {0,1,-1} locally nilpotent assignment which shows that T(2,1,3) is also nonnilpotent. We thus have the following result.

Theorem 4.4 *Among nilpotent trapezoids with lateral side 1, only those of the form T(3t, 1, 3t+1) are nilpotent. For t ≥1 these are composite.*

$\prod_4$ trapezoids with lateral side 2 fall into two categories as illustrated in Figure 4.15.

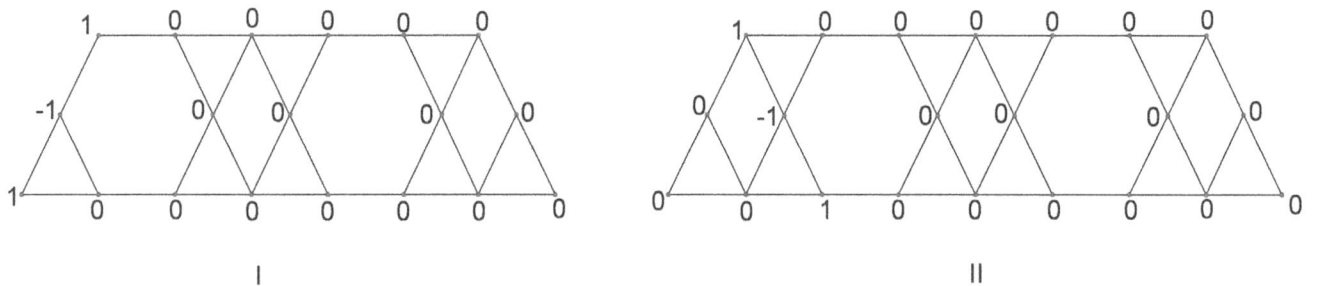

Figure 4.15

Both types of trapezoids are given a {0,1,-1} locally nilpotent assignment with all 0 tails. As a consequence each trapezoid of either type has a sum of 1. The conclusion is the following.

Theorem 4.5 *$\prod_4$ trapezoids with lateral side 2 are all nonnilpotent.*

82

4.5.2 $\prod_4$ Trapezoids with lateral side 3

Next we consider $\prod_4$ trapezoids with lateral side 3. In Type(I) the left corner is occupied by a nilpotent Δ_3 as in Figure 4.16.

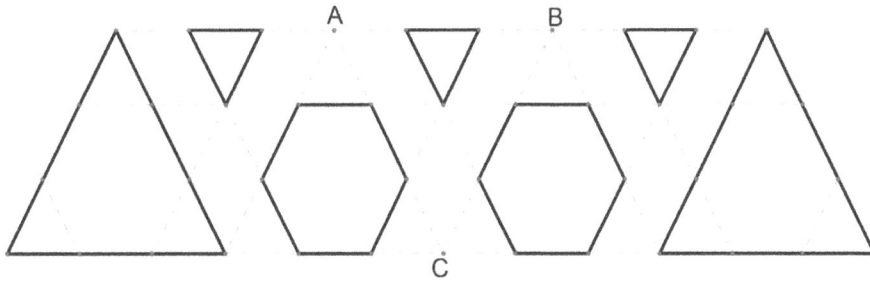

Figure 4.16

This Figure shows a composite T(9,3,12) trapezoid with factors consisting of two nilpotent Δ_3 triangles which flank a composite T(4,3,7) trapezoid. Note that by Theorem 4.1(V) the points A,B,C have a 0 sum. Also contained in Figure 4.16 is a composite P(3,8) parallelogram which can be repeatedly annexed to show that trapezoid T(9t,3,9t+3) is composite for every t ≥1.

These are the only nilpotent Type(I) trapezoids, (with lateral side 3), as we will now demonstrate. We will employ two {0,1,-1} patterns as in Figure 4.17.

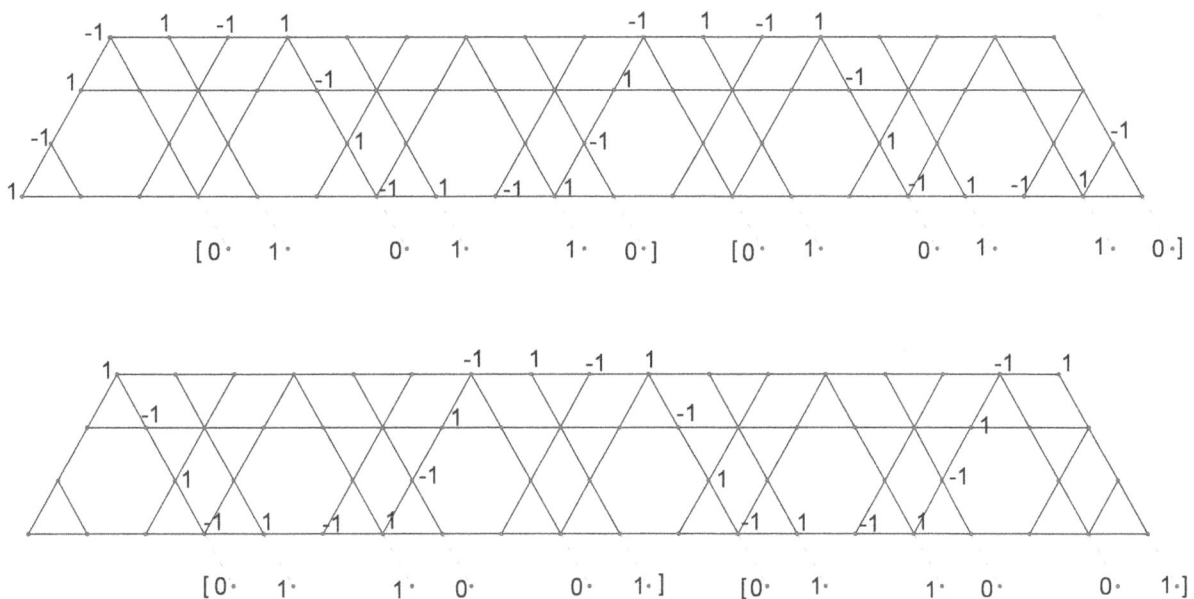

Figure 4.17

83

In this Figure unlabeled points have 0 value in the pattern. If the pattern were extended to the entire Π_4 plane, a lattice composed of H_3 hexagons would be obtained, each hexagon with a boundary consisting of alternating 0's and 1's, and an interior consisting entirely of 0's. We call this the H_3 *pattern*. The sums of the corresponding subtrapezoids are indicated below each trapezoid. The H_3 pattern is employed in two different versions in Figure 4.17, and wherever a nonzero value occurs as a subtrapezoid sum, the subtrapezoid must be nonnilpotent.

It is expected that the sequence of subtrapezoid sums is repetitive due to the regular construction of the Π_4 trapezoid as well as the regular nature of the pattern. It repeats in subsequences of length 6 as indicated in the Figure, the distance between repeated values being 9. From the top trapezoid in Figure 4.17 we can conclude that all Type(I) trapezoids with lower bases of one of the forms {9t, 4+9t, 7+9t} are nonnilpotent.

From the lower trapezoid in Figure 4.17 we can conclude that Type(I) trapezoids with lower bases of the forms {1+9t, 4+9t, 6+9t} are nonnilpotent. Lower bases which do not correspond to trapezoids have the form 2+3t, which is equivalent bases of any of the forms {2+9t, 5+9t, 8+9t}. From these observations, we can finally conclude the following.

Theorem 4.6 *Type(I) trapezoids with lateral side 3 are nilpotent iff they have the form T(9t, 3, 9t+3), t ≥1.*

A Type(II) trapezoid, (lateral side 3), is illustrated in Figure 4.18.

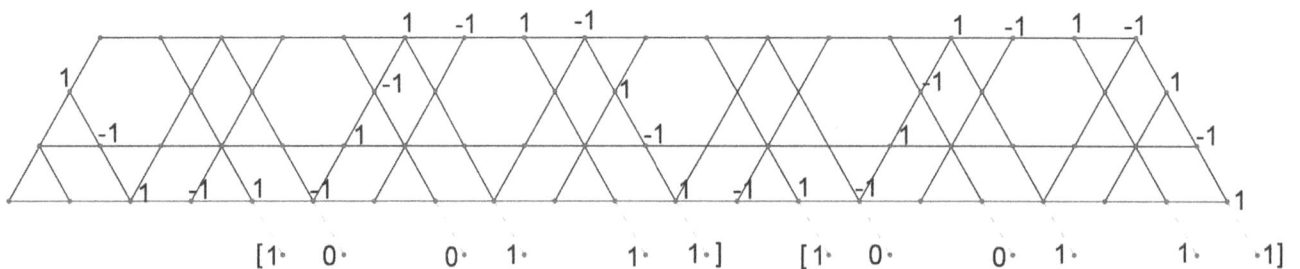

Figure 4.18

From this Figure we immediately obtain that Type(II) trapezoids with lower bases of the forms {1+9t, 2+9t, 4+9t, 5+9t, 8+9t} are nonnilpotent. Lower bases of the forms {3+9t, 6+9t, 9t} do not yield trapezoids. This leaves only trapezoids with lower bases of the form {7+9t} to consider. In Figure 4.16 clip off the leftmost Δ_3, invert the figure and append another T4,3, 7) trapezoid to obtain a composite T(13,3,16) trapezoid. To this figure append any number of composite P(3,8) parallelograms to obtain all (nilpotent) trapezoids of the form T(4+9t, 3, 7+9t), t ≥0.

Type(III) and Type(IV) trapezoids with lateral side 3, have the initial H_1 component nonadjacent to the left side, as in Figure 4.19.

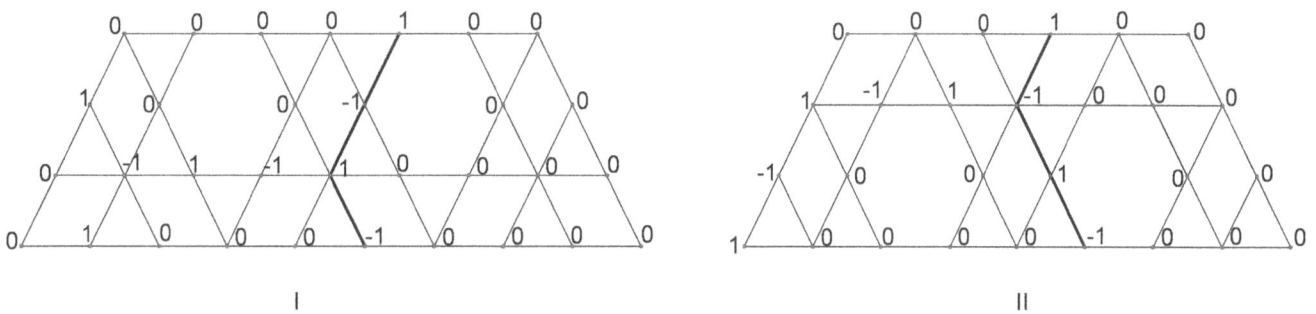

Figure 4.19

In Figure 4.19(I) the value of the locally nilpotent assignment to the left of the indicated broken line segment is 1, and the value to the right is endlessly 0. The conclusion is that no trapezoids of Type III are nilpotent. In Figure 4.9(II) it is easily seen that T(2,3,5) is a composite of two Δ_1's and a prime H(1,2,1,2,1,2) hexagon. But for larger bases an all 0 assignment is possible yielding a value of 1 for the remaining Type(IV) trapezoids. In summary we have the following.

Theorem 4.7 *A $\prod_4$ trapezoid with lateral side 3 is nilpotent if it is Type(I) with the form T(9t, 3, 9t+3), t ≥1, or it is Type(II) with the form T(4+9t, 3, 7+9t), t ≥0. All Type(III) trapezoids are nonnilpotent, and T(2,3,5) is the only nilpotent Type(IV) trapezoid. All of the nilpotent trapezoids are composite.*

4.5.3 $\prod_4$ trapezoids with lateral side 4

$\prod_4$ trapezoids with lateral side 4 occur in two Types. Type(I) is illustrated in Figure 4.20.

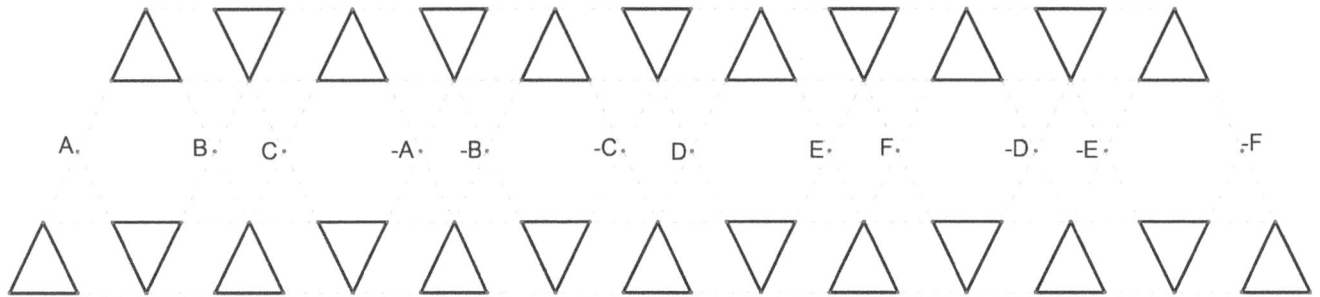

Figure 4.20

This Figure illustrates what we call a *simple 0-value dissection*. In this type of vertex dissection only nilpotent polygons, 0-sum point couples and 0-sum point triples are allowed. The Figure has a factorization consisting of a T(6,4,10) trapezoid and a P(4,8) parallelogram both of which are composite; each is an extension of a prime H(2,6,2,2,6,2) hexagon. The initial T(6,4,10) trapezoid can be extended ad infinitum using a series of P(4,8) parallelograms. We thus obtain the following result.

Theorem 4.8 *$\prod_4$ trapezoids of Type(I) and lateral side 4 are nilpotent iff they have the form T(6+9t, 4, 10+9t), t ≥0.*

The "only if" part of this Theorem can be proved using two H_3 patterns. In Figure 4.21 a Type(II) $\prod_4$ trapezoid with lateral side 4 is illustrated with an H_3 pattern.

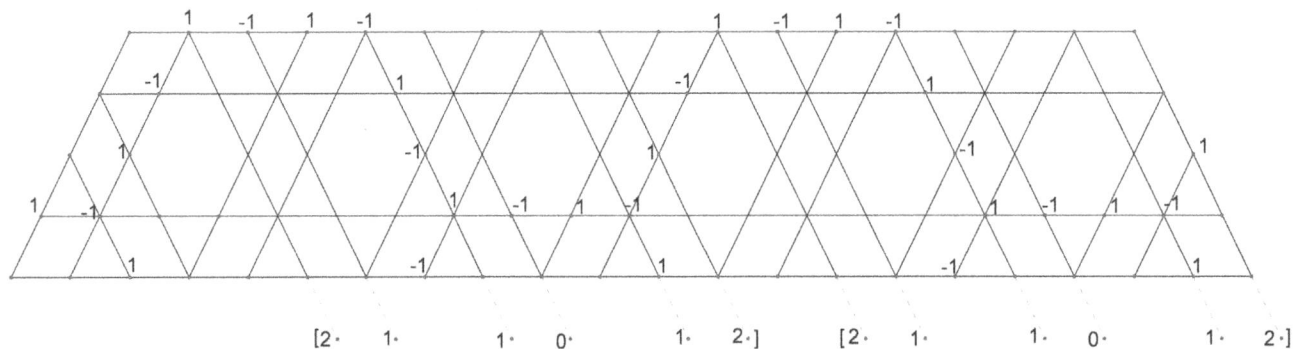

Figure 4.21

86

The H_3 pattern in Figure 4.21 shows that Type(II) trapezoids with bases of the forms $\{2+9t, 5+9t, 6+9t, 8+9t\}$ are nonnilpotent. By shifting this H_3 pattern 6 units to the left a pattern is obtained which shows that Type(II) trapezoids with bases of the form $\{9t\}$ are also nonnilpotent. This accounts for all possible trapezoid bases, and so we can conclude that $\prod_4$ trapezoids of Type(II) are all nonnilpotent. Thus we can strengthen Theorem 4.8:

Theorem 4.9 *A $\prod_4$ trapezoid with lateral side 4 is nilpotent iff it is Type(I) with the form T(6+9t, 4, 10+9t), t ≥0. Of these, only T(6,4,10) is prime.*

4.5.4 $\prod_4$ trapezoids with lateral side 5

There are two Types of $\prod_4$ trapezoid with lateral side 5. Type(I) is illustrated in Figure 4.22.

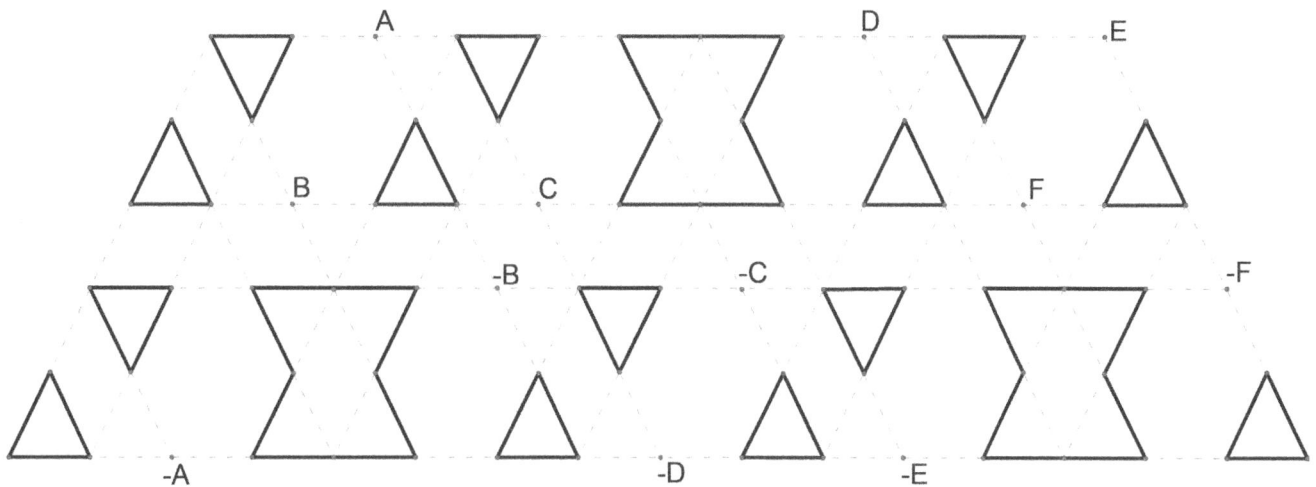

Figure 4.22

This Figure illustrates a simple 0-value dissection of a T(11,5,16) trapezoid. This trapezoid has a factorization consisting of a prime T(2,5,7) trapezoid and a nilpotent P(5,8) parallelogram. Any number of copies of this parallelogram can be appended on the right hand side with the result that every trapezoid of the form T(2+9t, 5, 7+9t) is nilpotent. H_3 patterns can be used to show these are the only nilpotent Type(I) trapezoids.

A Type(II) $\prod_4$ trapezoid is illustrated in Figure 4.23.

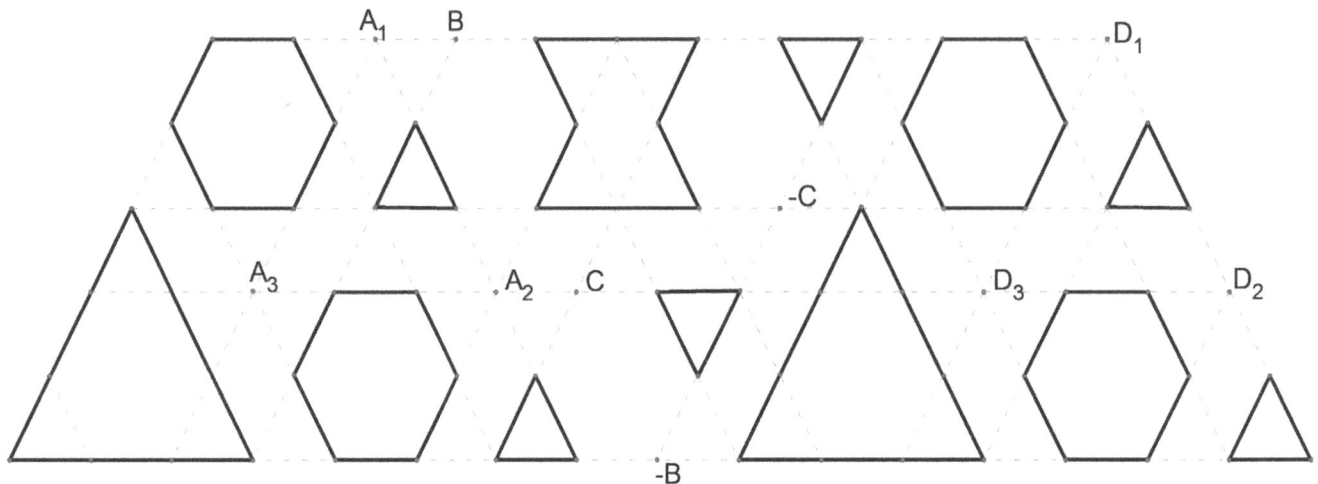

Figure 4.23

A simple 0-value dissection of a T(11,5,16) is depicted which extends a composite T(2,5,7) trapezoid. In Figure 4.23 we also see that T(11,5,16) is composed of a P(5,8) factor and a T(2,5,7) factor. Note that the T(2,5,7) has a factorization consisting of a Δ_3, a Δ_1 and a prime H(2,3,2,1,4,1). The section of the T(11,5,16) trapezoid which lies outside the initial T(2,5,7) can be appended repeatedly to obtain the result that the trapezoids T(2+9t, 5, 7+9t), t ≥0, are nilpotent. It can be shown using a single H_3 pattern that these are the only nilpotent Type(II) trapezoids with lateral side 5. We thus have the following result.

Theorem 4.10 *There are two Types of $\prod_4$ trapezoids with lateral side 5. For each Type only trapezoids of the form T(2+9t, 5, 7+9t), t ≥0 are nilpotent. All trapezoids of both Types are composite.*

4.5.5 $\prod_4$ trapezoids with lateral side 6

A Type(I) trapezoid T(15,6,21) with lateral side 6 is illustrated in Figure 4.24 with a simple 0-value dissection.

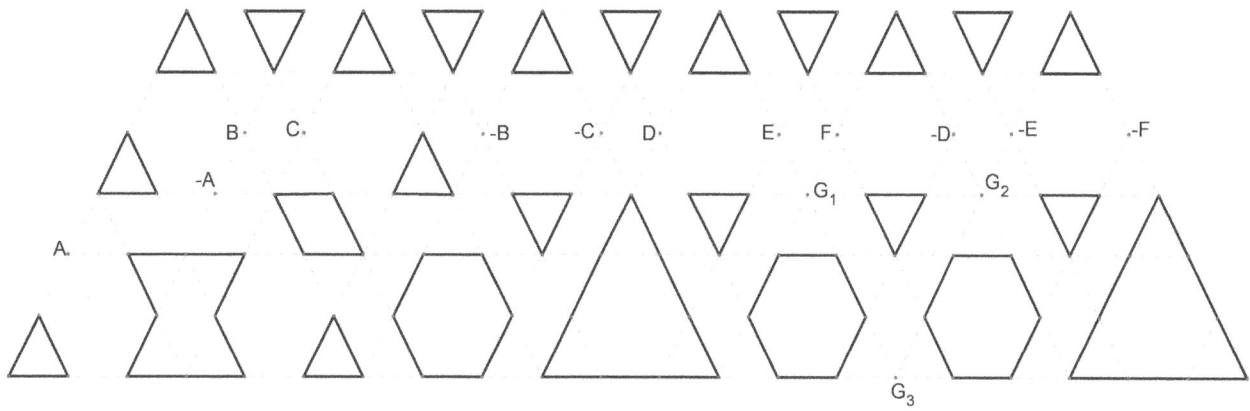

Figure 4.24

The Figure shows a factorization of T(15,6,21) into a T(6,6,12) trapezoid and a P(6,8) parallelogram both of which are composite. Viewed from right to left the Figure reveals a different simple 0-value dissection of a T(6,6,12) trapezoid along with a different simple 0-value dissection of P(6,8). Copies of the P(6,8) parallelogram can be appended at the right end to obtain the result that every Type(I) trapezoid of the form T(6+9t, 6, 12+9t), t ≥0 is nilpotent. A single H_3 pattern can be used to show that these are the only Type(I) trapezoids with lateral side 6.

A Type(II) trapezoid T(13,6,19) with lateral side 6 is illustrated with a simple 0-value dissection in Figure 4.25. This trapezoid has a factorization consisting of a composite T(4,6,10) and a composite P(6,8) parallelogram. Both of these structures contain a prime H(4,4,4,2,6,2) hexagon as a factor. Copies of P(6,8) can be appended at the right end creating nilpotent trapezoids of the form T(4+9t, 6, 10+9t), t ≥0.

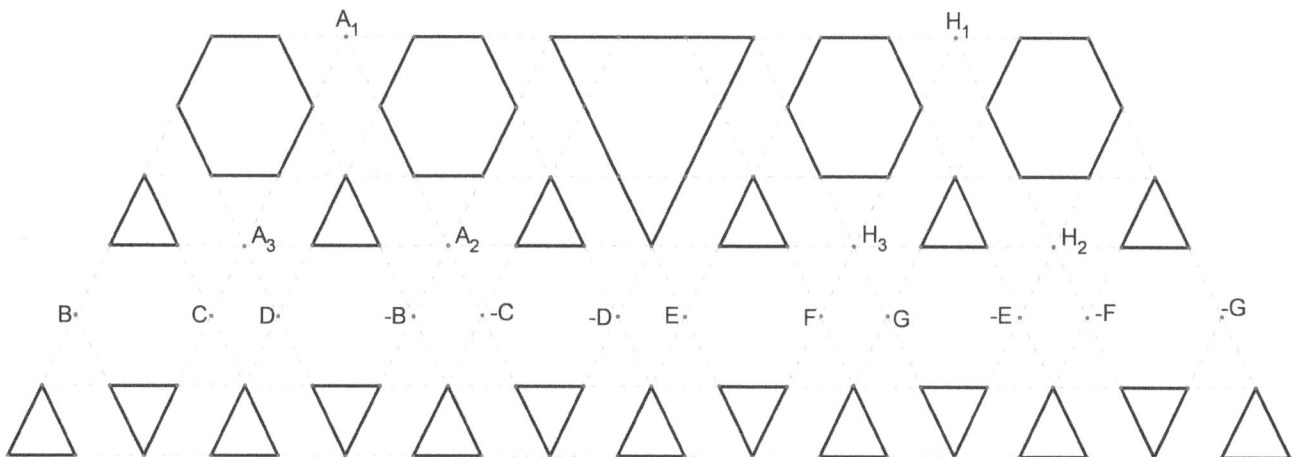

Figure 4.25

89

As it turns out the trapezoid T(2,6,8) is also nilpotent. This yields the construction illustrated in Figure 4.26.

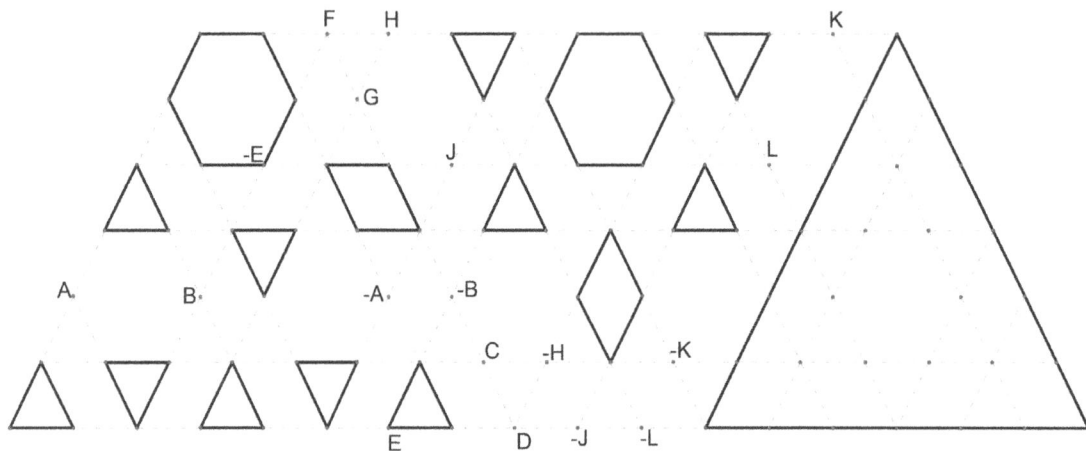

Figure 4.26

We could not find a simple 0-value dissection for T(2,6,8). Its nilpotence can be understood by considering the Figure. Note that C+D = E and F+G = -E, and consequently F+G+C+D = 0. The remainder of T(2,6,8) has a simple 0-value dissection as indicated in Figure 4.26. The section of Figure 4.26 which lies to the right of the initial T(2,6,8) trapezoid completes a nilpotent T(11,6,17). This section can be appended repeatedly to the right side to obtain that every Type(II) trapezoid of the form T(2+9t, 6, 8+9t), t ≥0, is nilpotent. An H_3 pattern can be used to show that there are no further Type(II) trapezoids with lateral side 6.

A Type(III) trapezoid T(11,6,17) is illustrated in Figure 4.27.

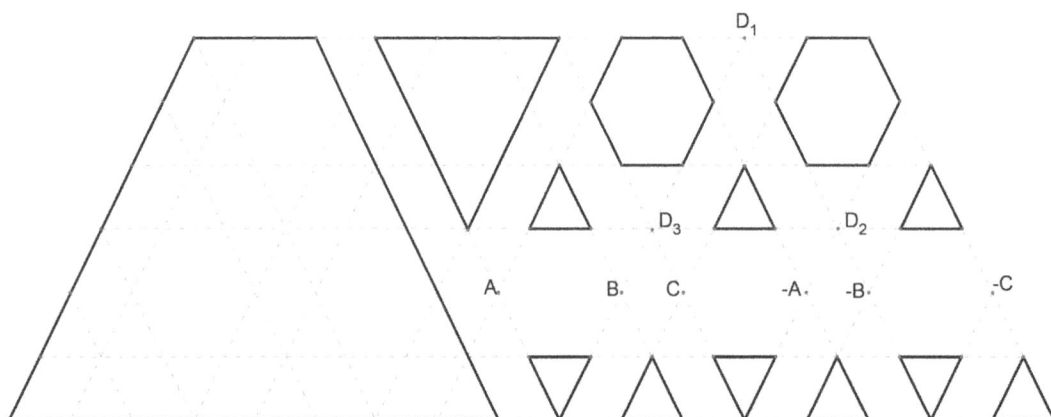

Figure 4.27

In this Figure the initial T(2,6,8) is just a reflection of the nilpotent T(2,6,8) in Figure 4.26. The remainder is a nilpotent P(6,8) parallelogram. This parallelogram can be appended indefinitely on the right hand side with the consequence that every Type(III) trapezoid of the form T(2+9t, 6, 8+9t), t ≥0 is nilpotent.

A second possibility for a Type(III) trapezoid with lateral side 6 is illustrated in Figure 4.28. A simple 0-value dissection is illustrated which features a nilpotent extension of a nilpotent Δ_6. The extension can be repeated ad infinitum to obtain the result that all Type(III) trapezoids of the form T(9t, 6, 6+9t), t ≥1, are nilpotent. A single H_3 pattern can be found which shows that besides the trapezoids indicated in Figures 4.27 and 4.28, there are no further nilpotent Type(III) trapezoids.

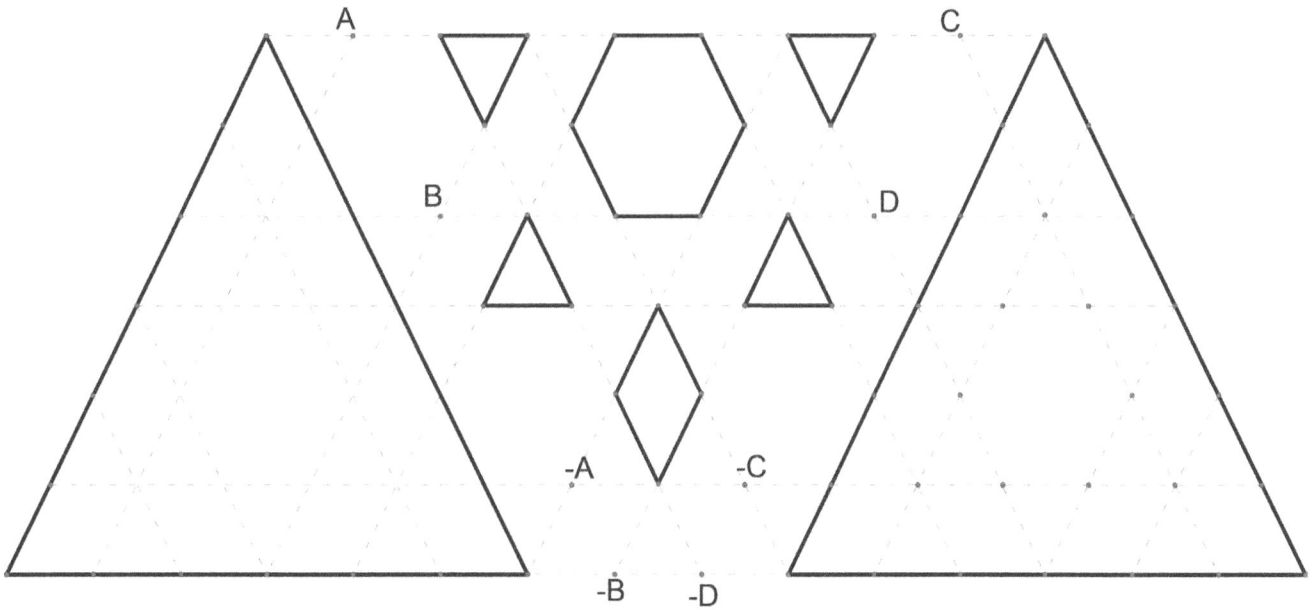

Figure 4.28

The last kind of $\prod_4$ trapezoid with lateral side 6 is Type(IV) illustrated in Figure 4.29.

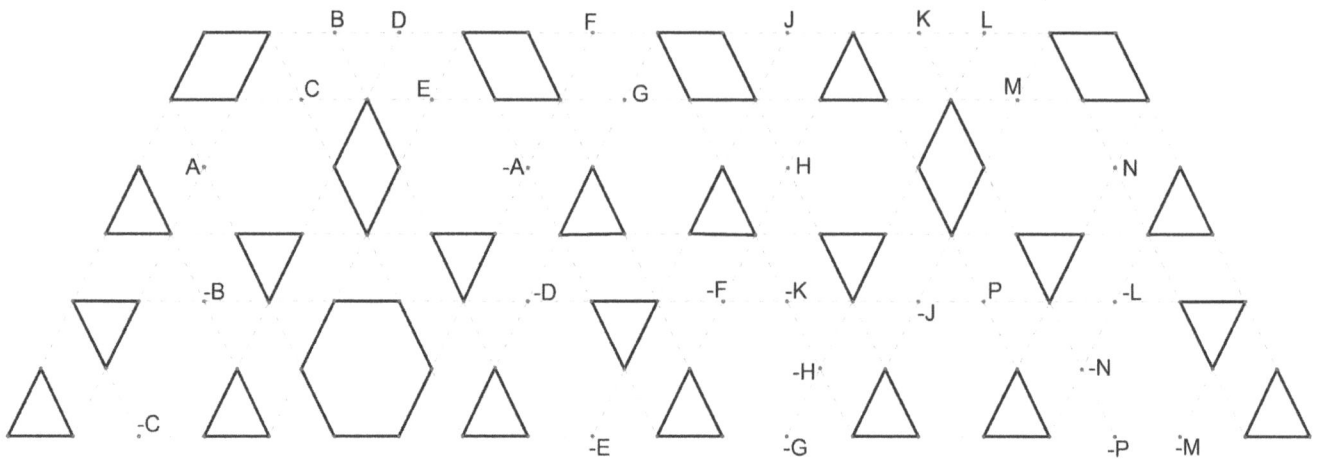

Figure 4.29

The Figure shows an extension of a nilpotent $T(5,6,11)$ to a nilpotent $T(14,6,20)$. The extension can be extended ad infinitum to obtain the result that Type(IV) trapezoids of the form $T(5+9t, 6, 11+9t)$, $t \geq 0$ are all nilpotent. A single H_3 pattern can be found which shows that all remaining Type(IV) trapezoids with lateral side 6 are nonnilpotent. The results regarding Π_4 trapezoids with lateral side 6 are summarized in the following.

Theorem 4.11 *A Π_4 trapezoid with lateral side 6 is nilpotent iff (i) it is Type(I) with the form $T(6+9t, 6, 12+9t)$, $t \geq 0$, or (ii) it is Type(II) with either of the forms $T(4+9t, 6, 10+9t)$, $t \geq 0$, $T(2+9t, 6, 8+9t)$, or (iii) it is Type(III) and has one of the forms $T(2+9t, 6, 8+9t)$, $t \geq 0$, $T(9t, 6, 6+9t)$, $t \geq 1$, or (iv) it is Type(IV) with the form $T(5+9t, 6, 11+9t)$, $t \geq 0$.*

At this point we can see that nilpotent Π_4 trapezoids can be put into three Classes: Class(I) consists of trapezoids which have a factorization consisting of a nilpotent triangle and a nilpotent parallelogram; Class(II) contains Π_4 trapezoids which have a factorization consisting of an initial nilpotent trapezoid and a nilpotent parallelogram; Class(III) which has an initial nilpotent trapezoid followed by an extension; Class(IV) which has an initial nilpotent triangle followed by an extension. In all cases the extension adds exactly 9 units to the bases. Since nilpotent parallelograms play an important part in the analysis of Π_4 trapezoids we pause to study parallelograms.

4.6 $\prod_4$ parallelograms

4.6.1 $\prod_4$ parallelograms of the form P(a,b) with a = 1, 2, 3

In case a = 1, it is easily seen that parallelograms of the form P(1,2+3t), t ≥0 are the only nilpotent ones, all of which are composite. It is also easy to see from Figure 4.15, that all parallelograms of the form P(2, m), m ≥2 are nonnilpotent. It is helpful to observe that trapezoids and parallelograms may be typed in the same way since the initial configurations are identical. Thus we may regard the P(3,8) parallelogram in Figure 4.26 as Type(I). By appending copies of this parallelogram at the right end, we obtain nilpotent parallelograms of the form P(3, 8+9t). These are the only Type(I) nilpotent parallelograms with side 3, as we can show using the locally nilpotent assignment illustrated in Figure 4.30. This is not a pattern since it cannot be extended to cover the entire $\prod_4$ plane.

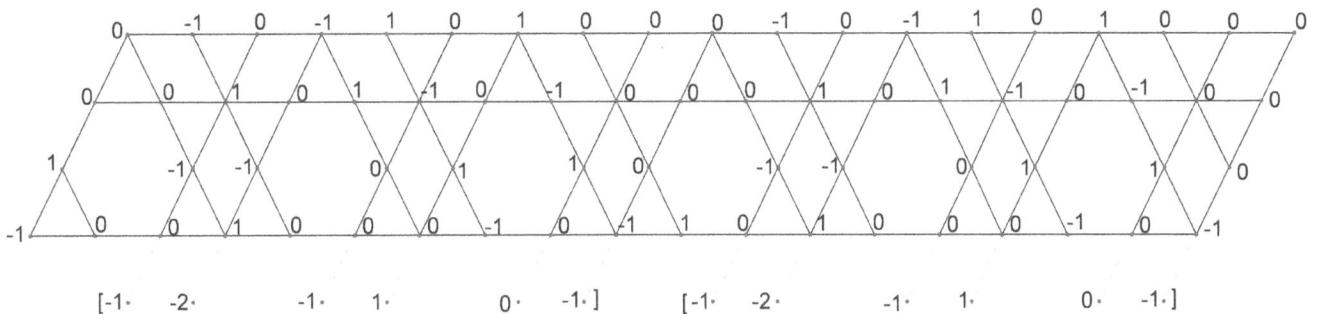

Figure 4.30

This assignment immediately eliminates the possibility of nilpotent parallelograms of Type(I) with side 3, besides the above mentioned. In Figure 4.31 a simple 0-value dissection of P(3,8) in a Type(II) parallelogram is illustrated. Copies of this parallelogram can be appended on the right side to obtain that P(3,8+9t) is nilpotent for t ≥0.

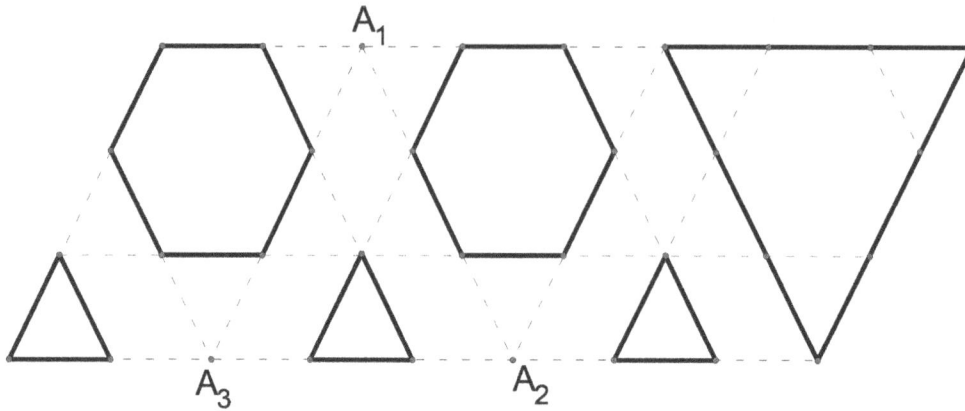

Figure 4.31

In Figure 4.32 a {0,1,-1} pattern is used to show that the above mentioned parallelograms are the only nilpotent ones of Type(II). We call this pattern the *diamond curtain* due to the way the 1's and -1's are situated.

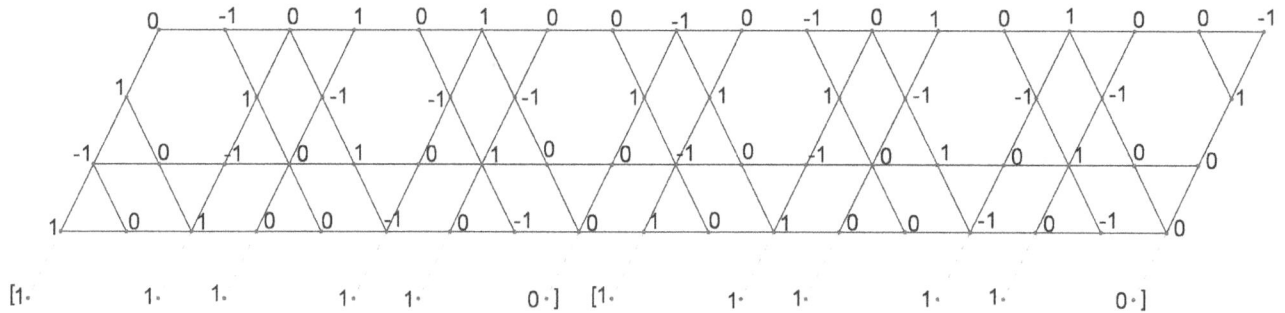

Figure 4.32

The {0,1,-1} pattern in Figure 4.19(I) can be used to show that there are no nilpotent Type(III) parallelograms with side 3. The locally nilpotent {0,1,-1} assignment illustrated in Figure 4.33 shows that there are no Type(IV) nilpotent parallelograms with side 3. This assignment cannot be extended to a pattern.

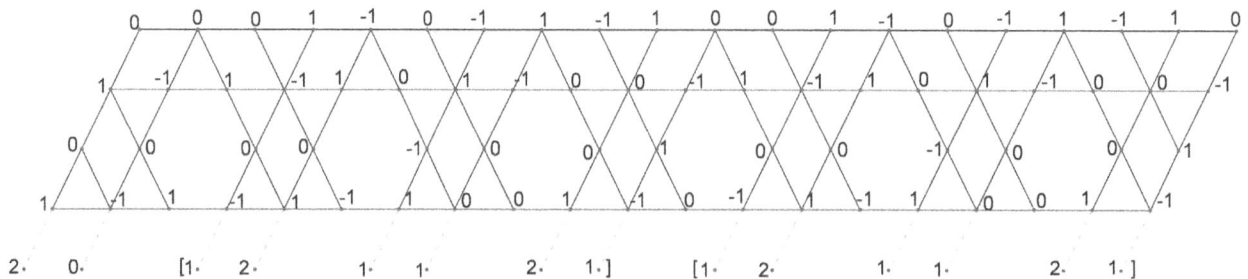

Figure 4.33

The results regarding nilpotent parallelograms with lateral side 3 is summarized in the following.

Theorem 4.12 $\prod_4$ *parallelograms with side 3 are nilpotent iff they are Type(I) or Type(II) and have the form P(3, 8+9t), t ≥0.*

4.6.2 $\prod_4$ parallelograms with lateral sides 4, 5, 6

From Figure 4.20 it is easily seen that Type(I) parallelograms of the form P(4,8+9t), t ≥0, are nilpotent. The H_3 pattern in Figure 4.34 shows that no more nilpotent Type(I) parallelograms with side 4 exist.

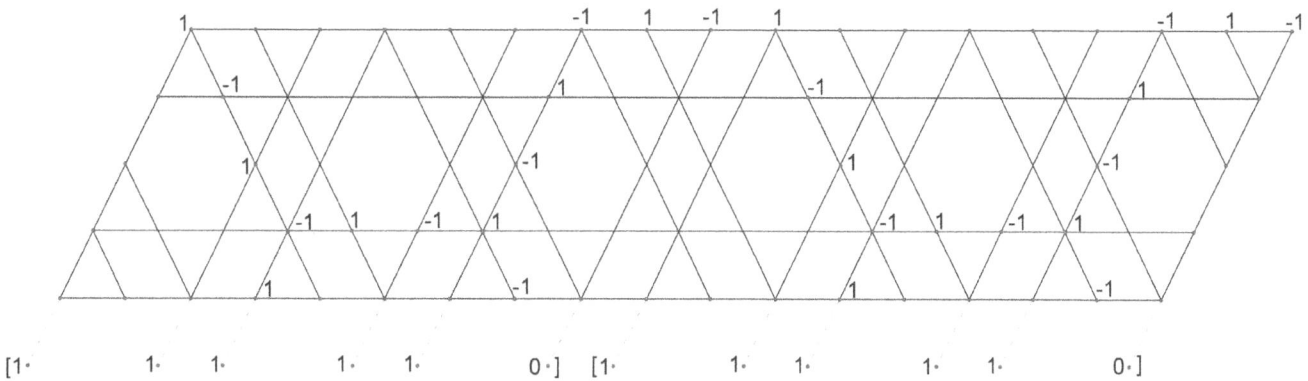

Figure 4.34

We leave it as an exercise to show that no Type(II) parallelograms with side 3 are nilpotent. There are only two Types of parallelograms with side 4, so we have the conclusion:

Theorem 4.13 *A $\prod_4$ parallelogram with side 4 is nilpotent iff it is Type(I) with the form P(4, 8+9t), t ≥0.*

In Figure 4.22 it can be seen that the $\prod_4$ parallelogram P(5,10) of Type(I) is nilpotent, and in fact the parallelograms P(5, 1+9t), t ≥0 are all nilpotent. In figure 4.23 we see that the Type(II) parallelograms P(5,8+9t), t ≥0 are all nilpotent. We leave it as an exercise to show that, besides these, there are no further nilpotent $\prod_4$ parallelograms with side 5. We thus have:

95

Theorem 4.14 *A $\prod_4$ parallelogram with side 5 is nilpotent iff it is Type(I) with the form P(5, 1+9t), t ≥0, or Type(II) with the form P(5, 8+9t), t ≥0.*

Figure 4.24 actually reveals both a Type(I) and a Type(II) parallelogram with side 6. If the Type(I) is rotated by 180 degrees, a Type(II) is obtained. Both are P(6,8) and can be extended to nilpotent parallelograms P(6, 8+9t), t ≥0. For further analysis we use an algebraic approach in Figure 4.35.

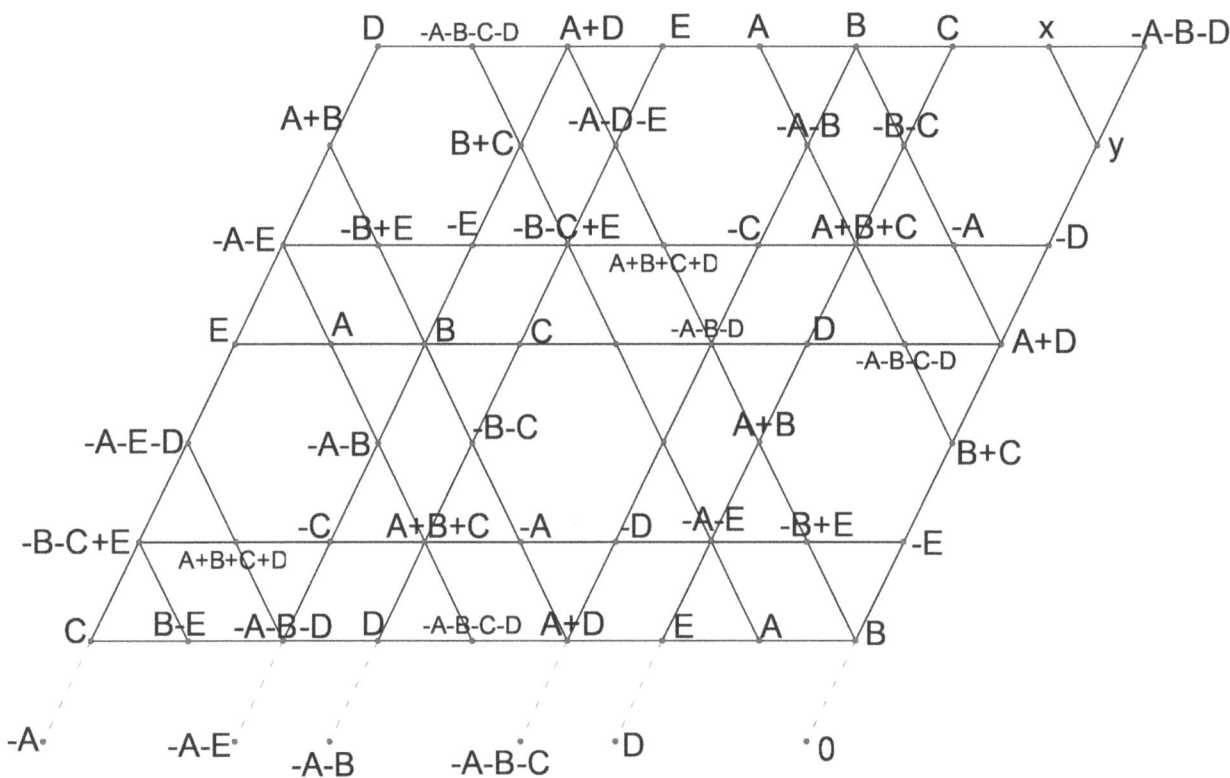

Figure 4.35

Now by setting A = D = 1, B = C = 0 we obtain the pattern in Figure 4.36. This is a {0,1,-1,2,-2} pattern in which the 1's and -1's occupy H_3 boundaries, though not alternating, and the 2's and -2's also occupy H_3 boundaries though not completely. The sums of the subparallelograms at the bottom of Figure 4.36 show that only the P(6,8) is nilpotent. If the pattern is continued to the right a repeat of these sums is obtained. The conclusion is that Type(I), and by rotation Type(II), $\prod_4$ parallelograms with side 6 are nilpotent iff they have the form P(6, 8+9t), t ≥0.

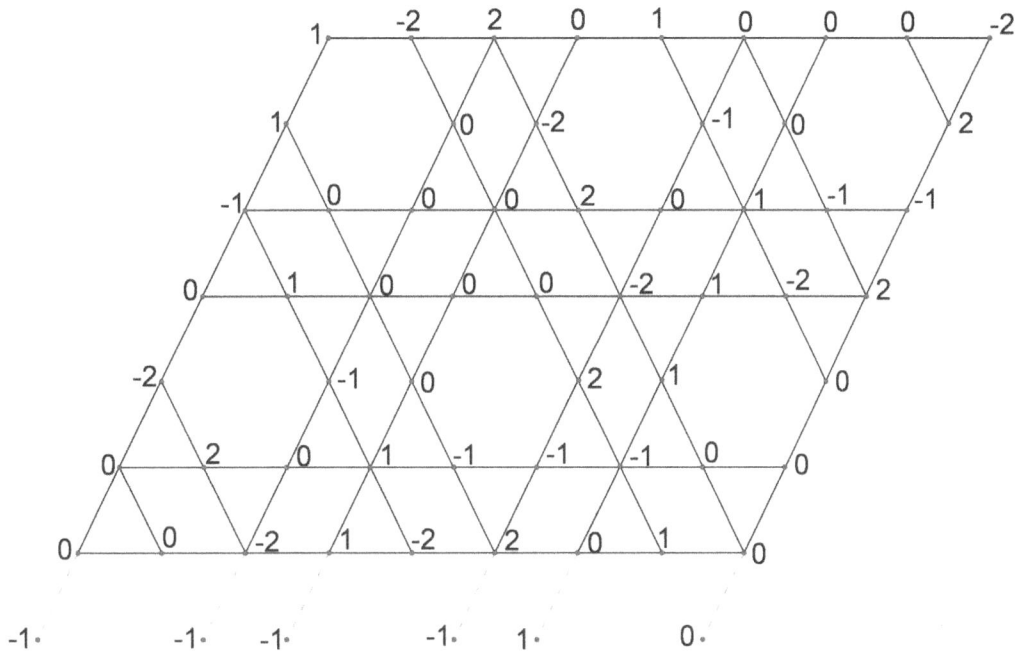

Figure 4.36

A nilpotent Type(III) rhomb R_6 with a P(6,8) extension is illustrated in Figure 4.37 with a simple 0-value dissection.

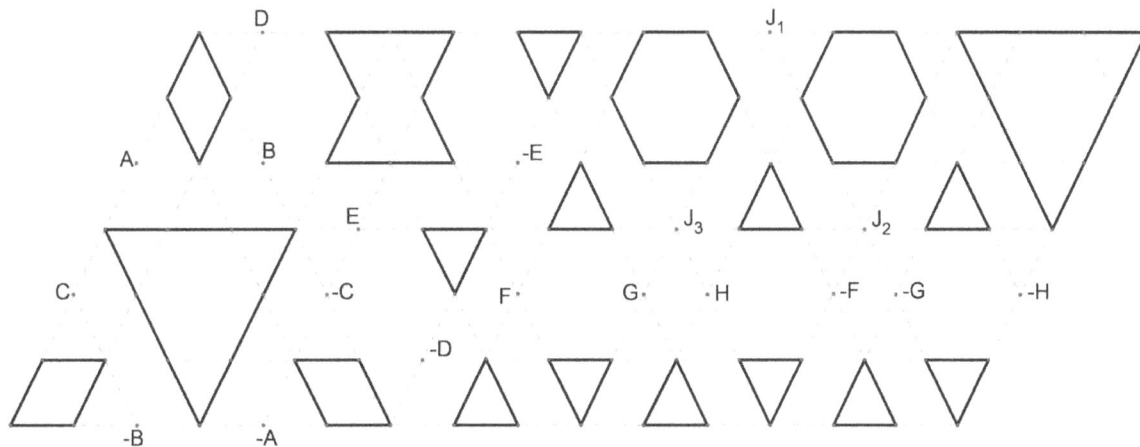

Figure 4.37

An algebraic analysis like the one illustrated in Figure 4.35 can be used to show that Type(III) parallelograms with side 6 are nilpotent iff they have the form P(6, 6+9t), t ≥0. To analyze Type(IV) parallelograms the following Lemma is useful.

Lemma 4.15 *In the H(1,4,2,1,4,2) hexagon in Figure 4.38 the indicated points x, y are equal in any locally nilpotent assignment.*

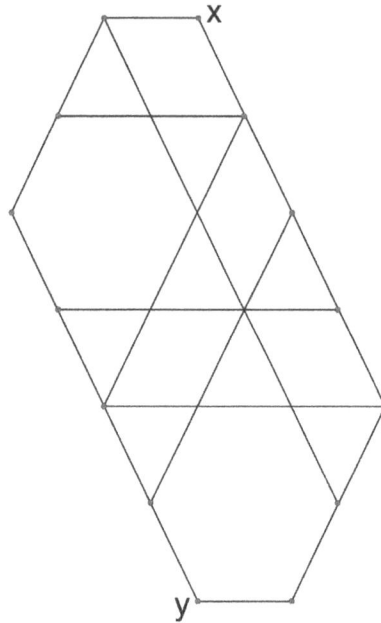

Figure 4.38

Lemma 4.38 and Theorem 4.1(vi) explain the sequence of values which occur on the lower base of the Type(IV) parallelogram in Figure 4.39. More significantly, the repeated subparallelogram sums indicated at the bottom of Figure 4.39, can be explained by the intervening Type(III) nilpotent R_6 rhombs. To understand further the sequence of sums consider the following Lemma.

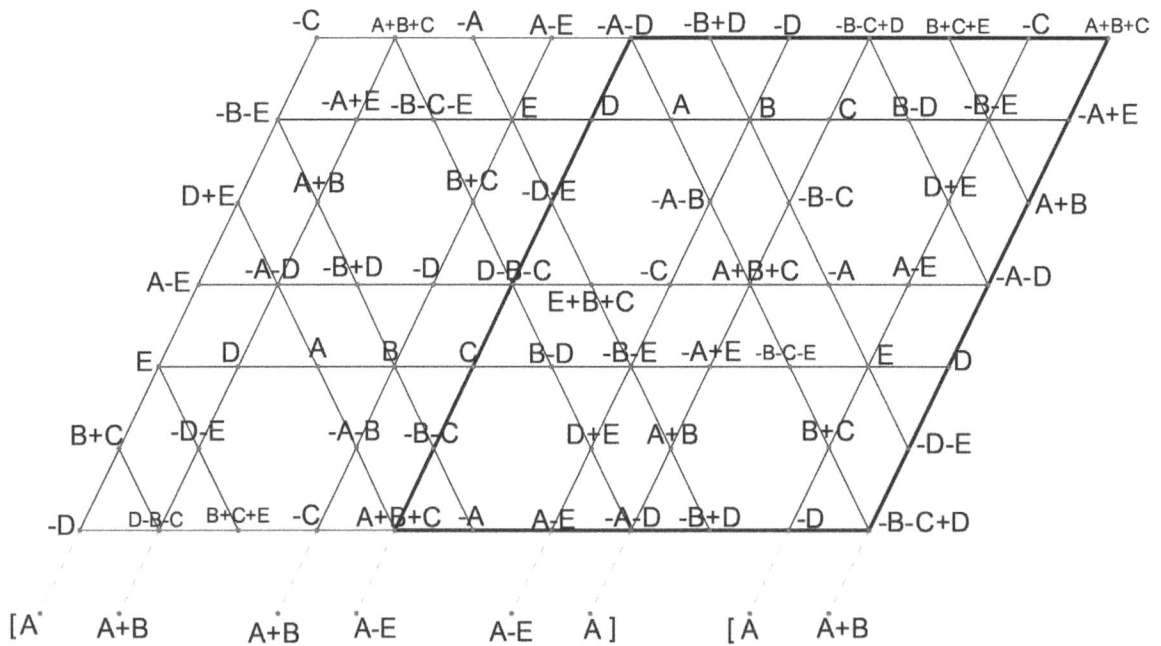

Figure 4.39

Lemma 4.16 *In the P(6,4) $\prod_4$ parallelogram in Figure 4.40, the sum of the indicated points is 0, i.e., x+y+z+u+v = 0.*

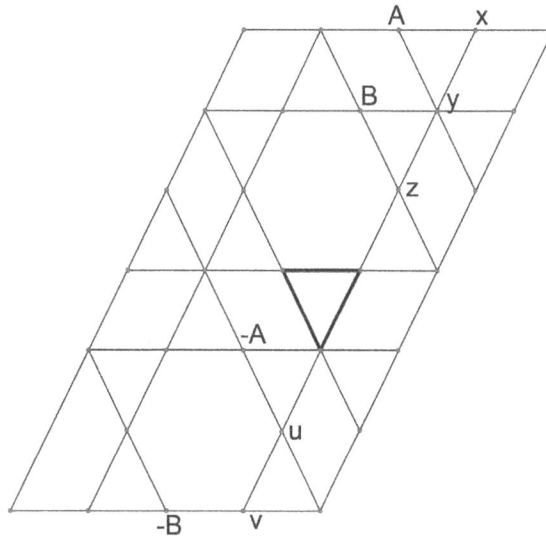

Figure 4.40

Lemma 4.16 explains why adjacent subparallelogram sums are the same. Integer values for A, B, E are easily chosen which make the subparallelogram sums listed in Figure 4.39 nonzero. Lemma 4.16 and the intervening nilpotent R_6 rhombs assure that all subparallelogram sums are nonzero, and consequently we can conclude that all Type(IV) parallelograms with side 6 are nonnilpotent. The following Theorem summarizes the results of this Section.

Theorem 4.17 *A $\prod_4$ parallelogram with side 6 is nilpotent iff (i) it is Type(I) or Type(II) and has the form P(6, 8+9t), t ≥0 or (ii) it is Type(III) and has the form P(6, 6+9t), t ≥0. All Type(IV) parallelograms with side 6 are nonnilpotent.*

4.7 Large $\prod_4$ parallelograms and trapezoids

We cannot give a full examination of parallelograms with side greater than 6 or trapezoids with lateral side greater than 6, so we will be content to illustrate how some larger nilpotent structures of these types can be constructed. We start by constructing nilpotent parallelograms with side 8 as in Figure 4.41. The Figure shows a nilpotent P(8,3) with an extension to a nilpotent P(8,6) and a further extension to a nilpotent P(8,9). Clearly this construction can be continued to obtain nilpotent parallelograms P(8,3t), t≥1.

In Figure 4.42 a construction of a nilpotent P(8,7) is illustrated. There is a clear extension to nilpotent parallelograms of the form P(8, 4+3t), t ≥0.

In Figure 4.43 a construction is indicated which shows by extension that parallelograms of the form P(8, 5+6t), t ≥0 are all nilpotent.

In Figure 4.44 a construction is given which shows that parallelograms of the form P(8, 8+6t), t ≥0 are nilpotent.

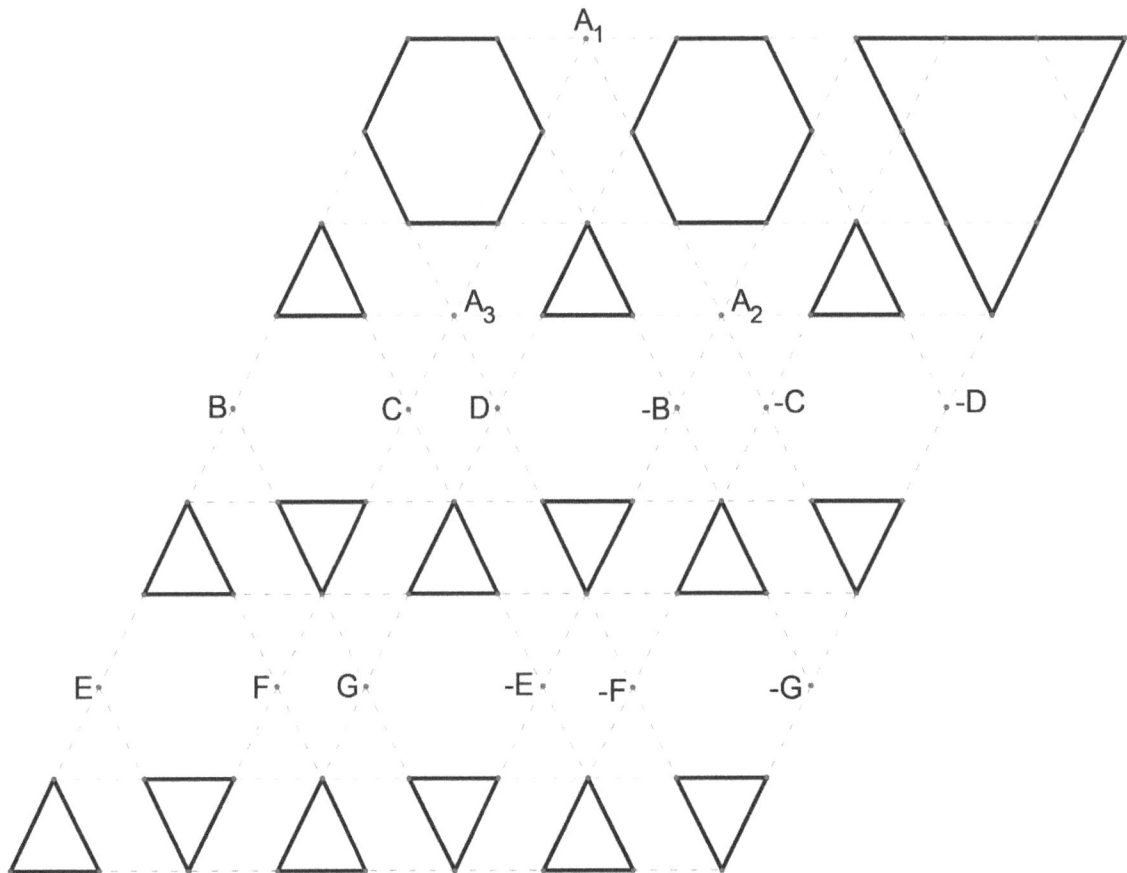

Figure 4.41

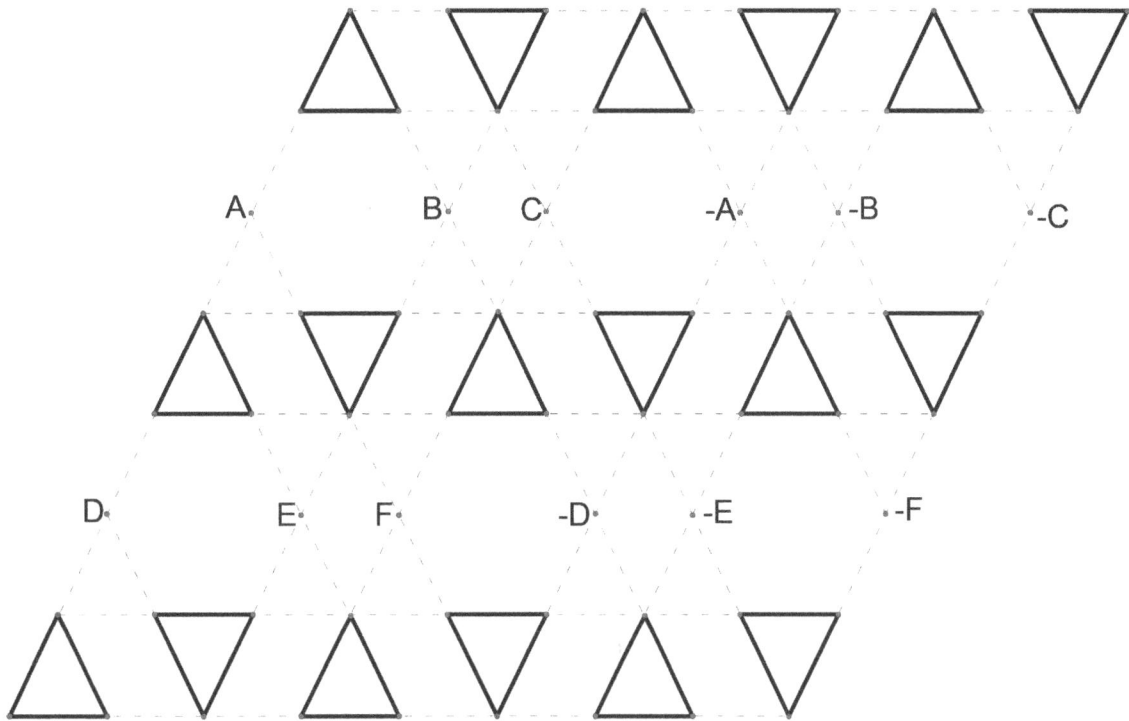

Figure 4.42

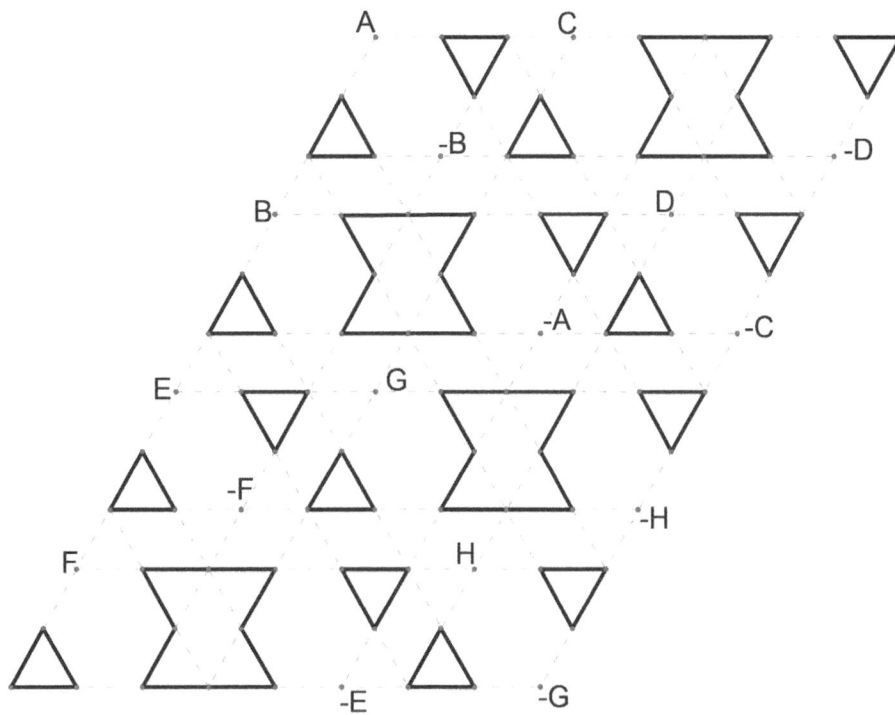

Figure 4.43

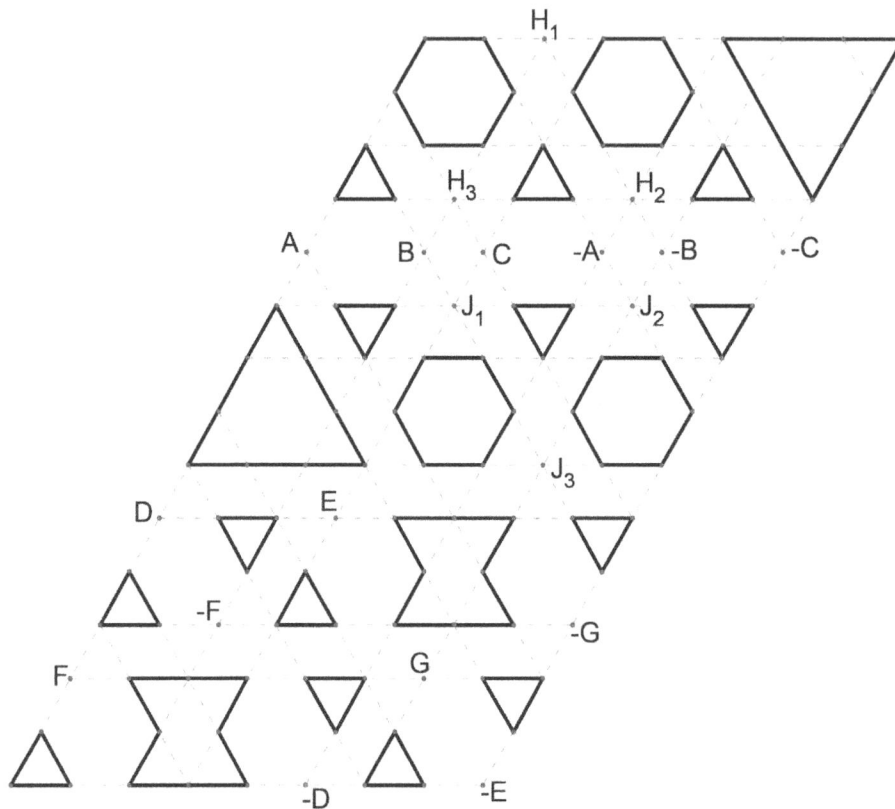

Figure 4.44

As a consequence of the foregoing constructions we can make the following general statement concerning $\prod_4$ parallelograms with side 8.

Theorem 4.18 *The $\prod_4$ parallelogram P(8, m) is nilpotent with m $\geq$3 if (i) it is Type(II) and has the form P(8, 3t), t $\geq$1, or (ii) it is Type(II) with the form P(8, 4+3t), t $\geq$0, or (iii) it is Type(I) and has the form P(8, 5+6t), t$\geq$0, or (iv) it is Type(I) with the form P(8, 8+6t), t $\geq$0.*

Thus there is a Type(I) or Type(II) nilpotent P(8, m) for every m $\geq$3. In Figure 4.45 a construction illustrates the nilpotence of R_7, P(7,16) and T(8,7,15).

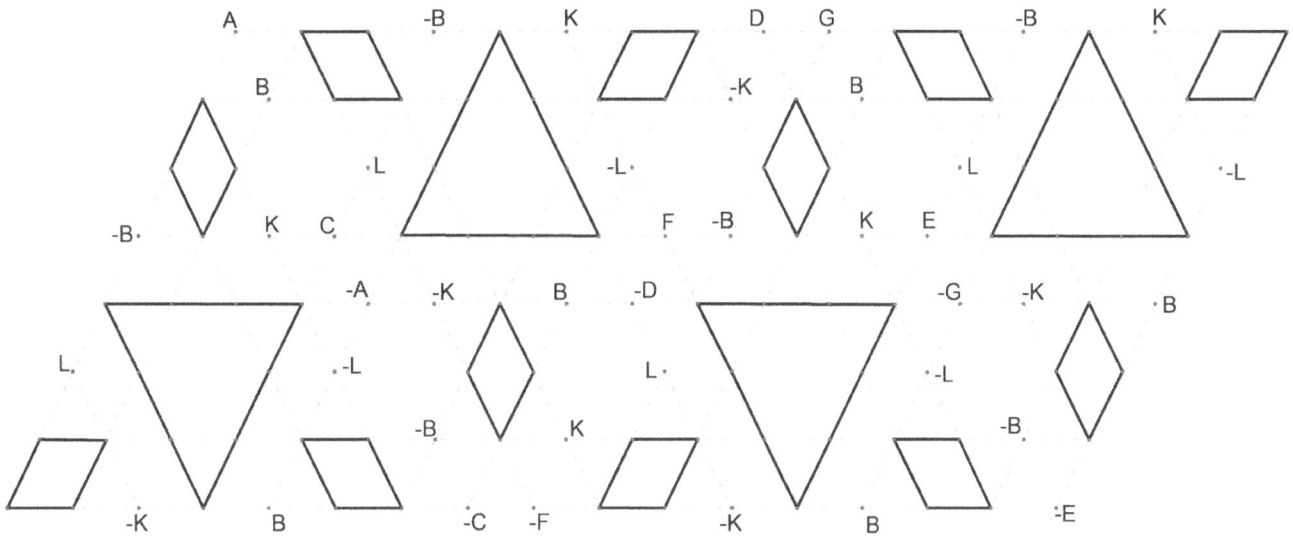

Figure 4.45

Note the repeated use of the nilpotent Δ_6 in this construction. By extension we obtain that Type(IV) parallelograms P(7, 7+9t), t ≥0, and trapezoids T(8+9t, 7, 15+9t), t ≥0 are nilpotent. In Figure 4.46 a nilpotent T(8,9,17) trapezoid is illustrated. It is composed of two nilpotent Δ_8 triangles and a nilpotent Δ_7 triangle. It is not possible to find a simple 0-value dissection for the Δ_8 triangles. Instead Lemma 4.19 below can be used to prove the nilpotence of these triangles.

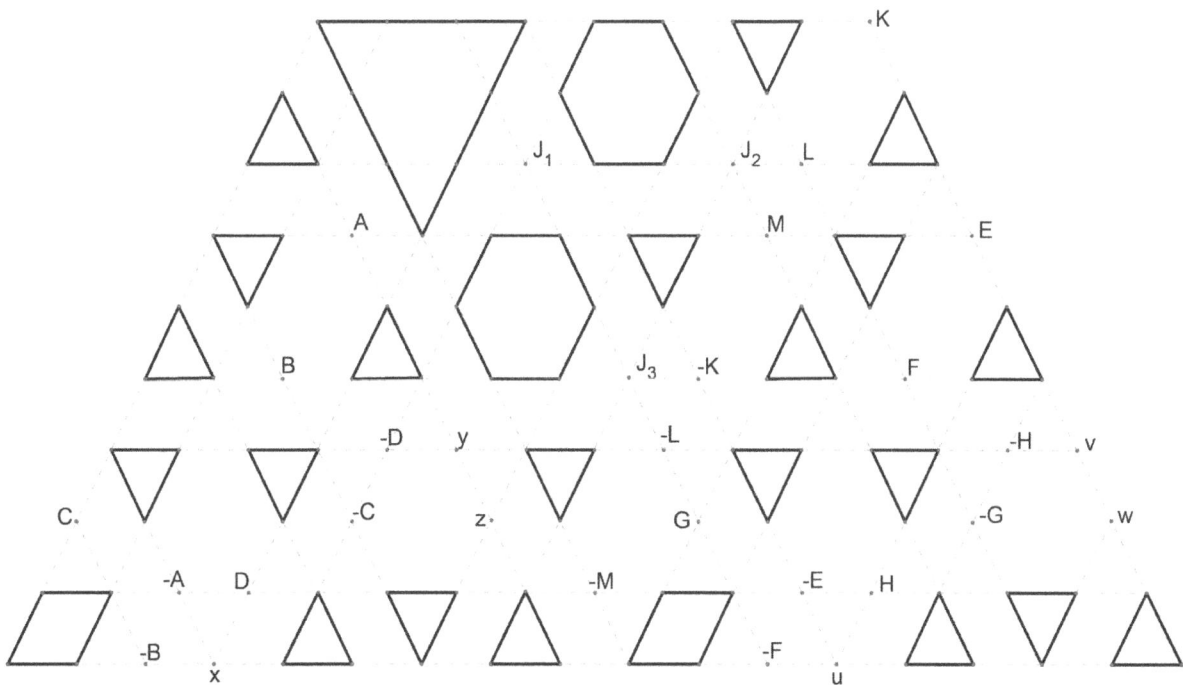

Figure 4.46

Lemma 4.19 *In the $\prod_4$ pentagon in Figure 4.47 the variables x, y, z satisfy x+y+z = 0.*

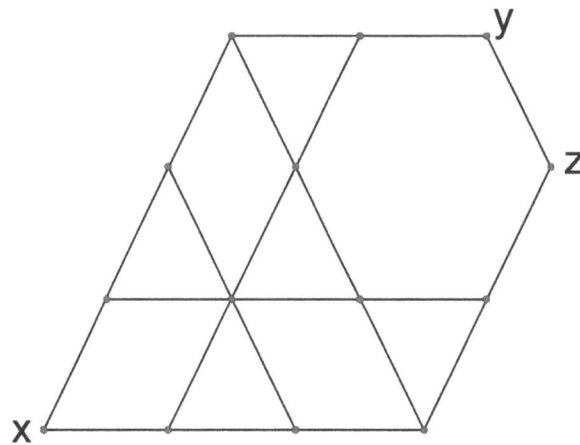

Figure 4.47

The construction in Figure 4.46 can be extended to obtain the nilpotence of trapezoids in the form T(8+9t, 9, 17+9t).

In Figure 4.48 the construction of a nilpotent P(10,20) is illustrated.

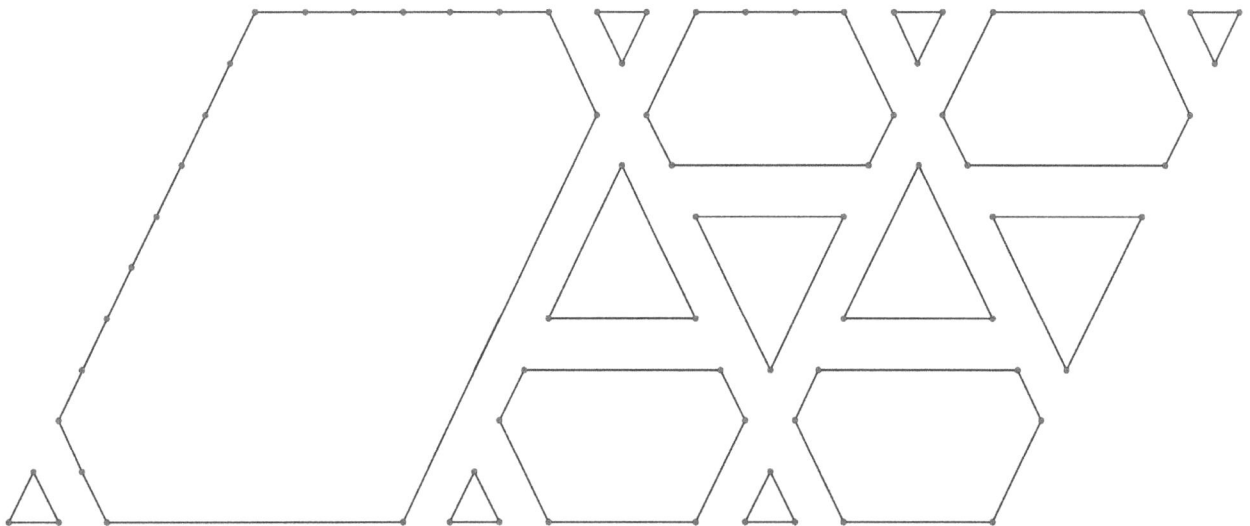

Figure 4.48

Note that a nilpotent P(10,14) occurs as a factor. Moreover the construction can be extended indefinitely by appending copies of the P(5,10) and/or the P(8,10) parallelograms.

4.8 Parallelogram conglomerates

A nilpotent $\prod_4$ parallelogram which has a dissection into smaller nilpotent parallelograms will be called a *parallelogram conglomerate*. The construction in Figure 4.48 and its extensions are thus parallelogram conglomerates. The parallelograms in Figures 4.37, 4.43 and 4.44 are also examples of parallelogram conglomerates. In Figure 4.49 an R_{15} is shown to be a conglomerate of P(6,15) and P(8,15). In Figure 4.50 a nilpotent R_{17} rhomb is illustrated as a conglomerate of two nilpotent P(8,17) parallelograms. Note that various dissections are possible whose main factors are nilpotent H_4 hexagons and Δ_3 triangles. A simple 0-value dissection of H_4 is given in Figure 4.51.

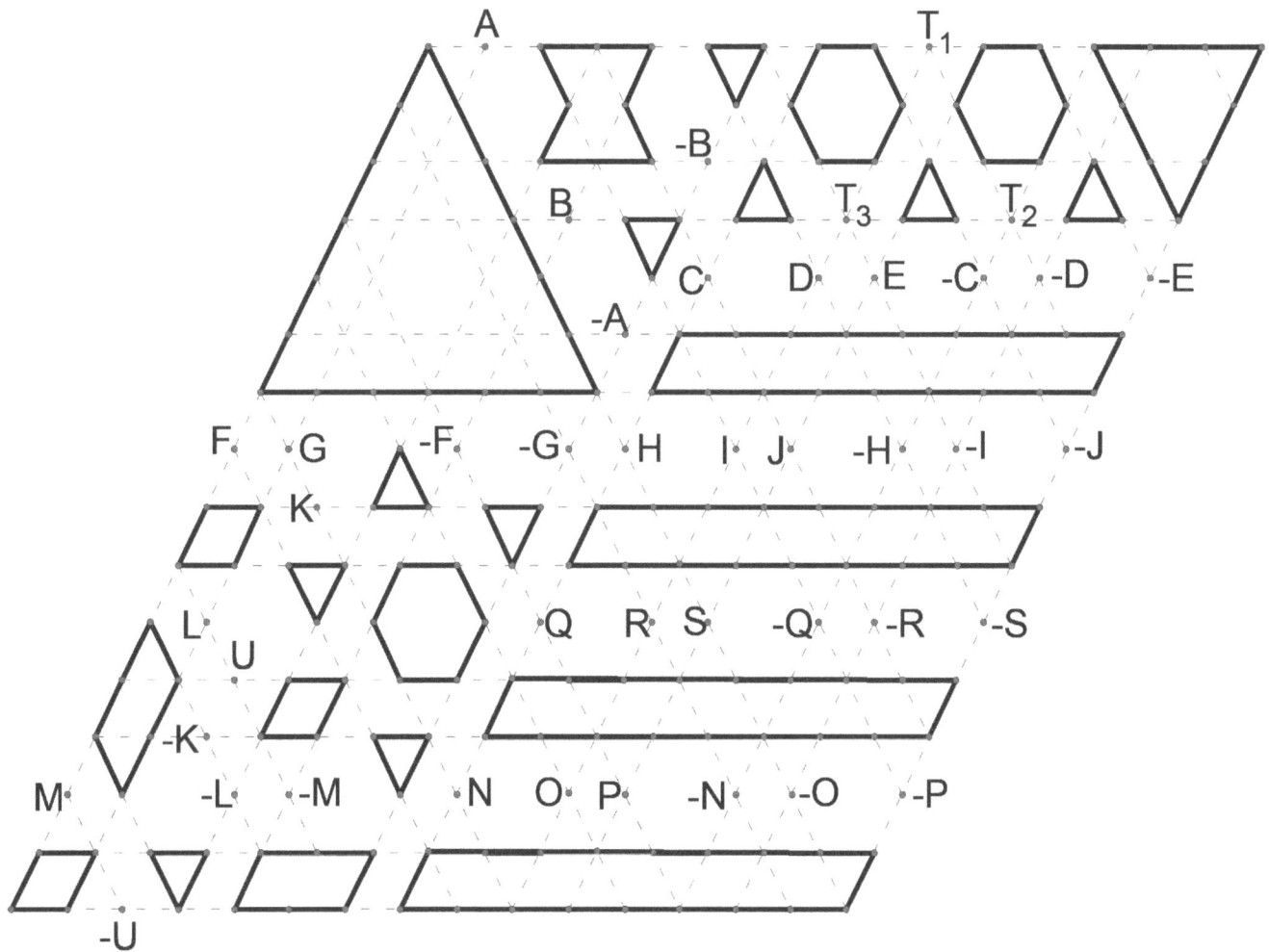

Figure 4.49

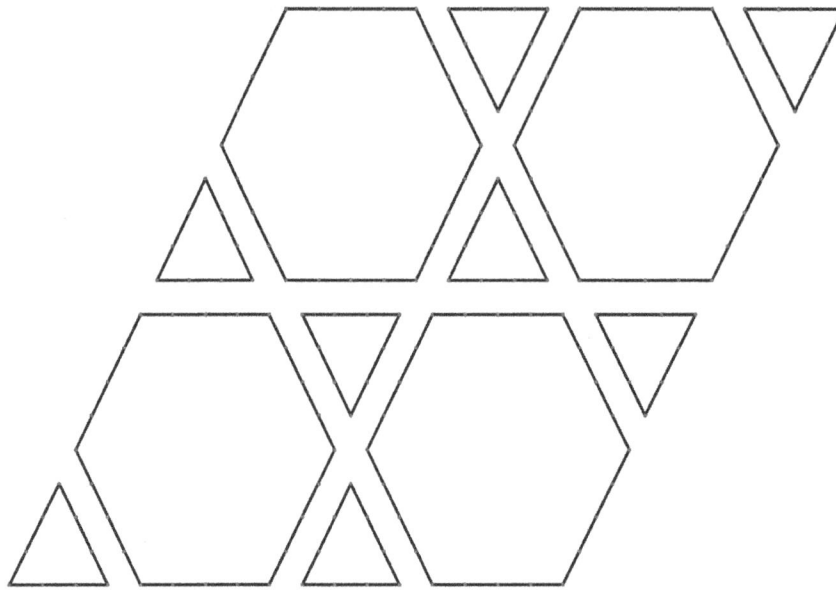

Figure 4.50

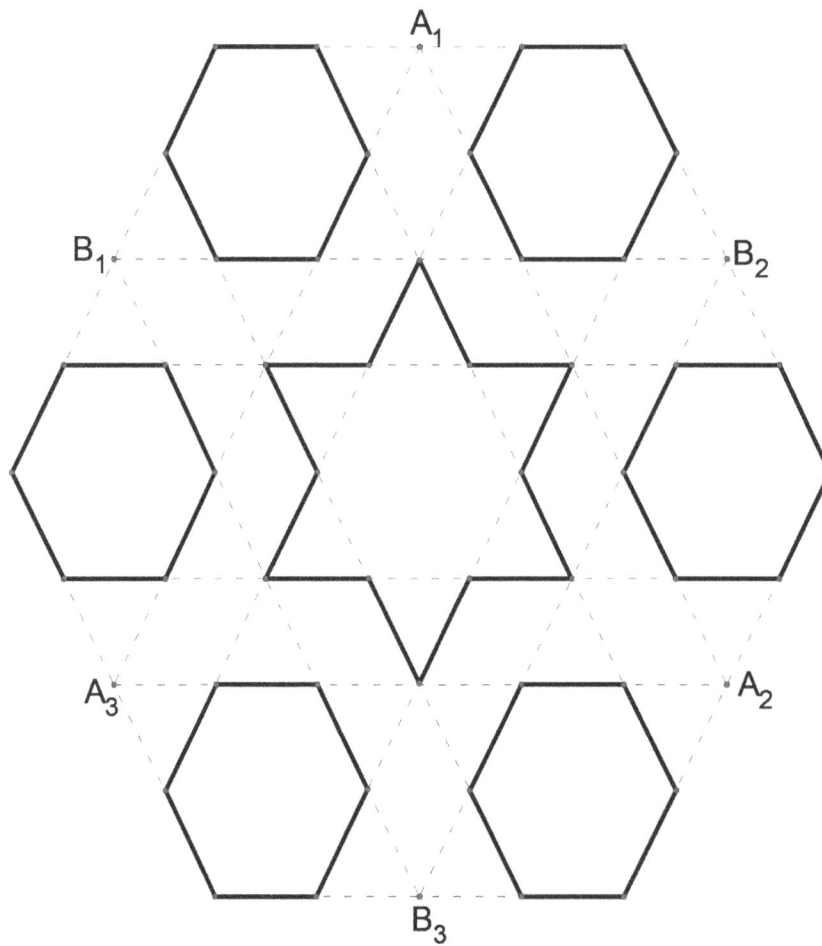

Figure 4.51

In Figure 4.52 a nilpotent R_{16} rhomb is constructed with a simple 0-value dissection. The main ingredient of this dissection is the nilpotent Δ_6 triangle. We have now constructed the nilpotent rhombs R_6, R_7, R_8, R_{15}, R_{16} and R_{17}. These can be extended to obtain nilpotent rhombs of the forms R_{6+9t}, R_{7+9t}, R_{8+9t}, $t \geq 0$. We leave it as an open question as to whether these constitute the full class of nilpotent $\prod_4$ rhombs. In Figure 4.53 we indicate a method which might be used to show that a given $\prod_4$ rhomb is *not* nilpotent.

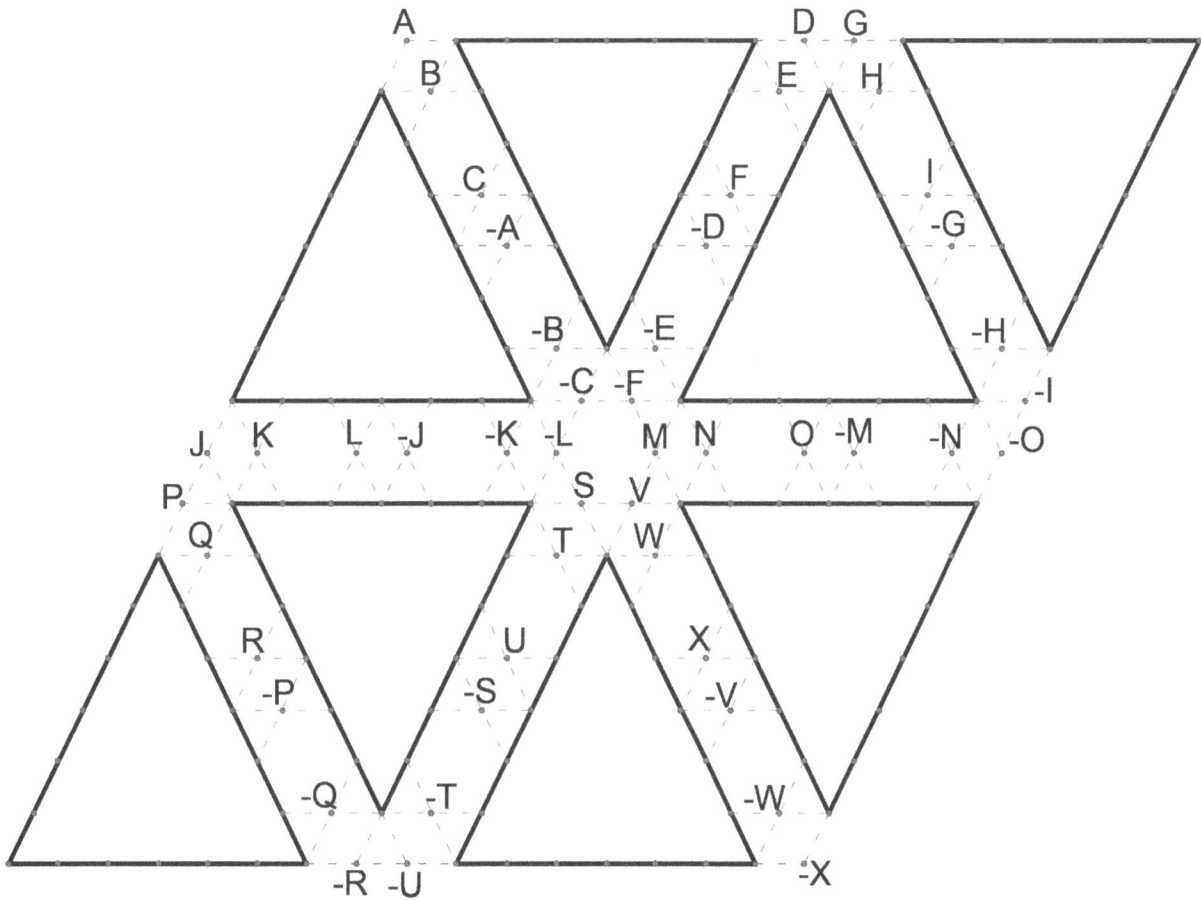

Figure 4.52

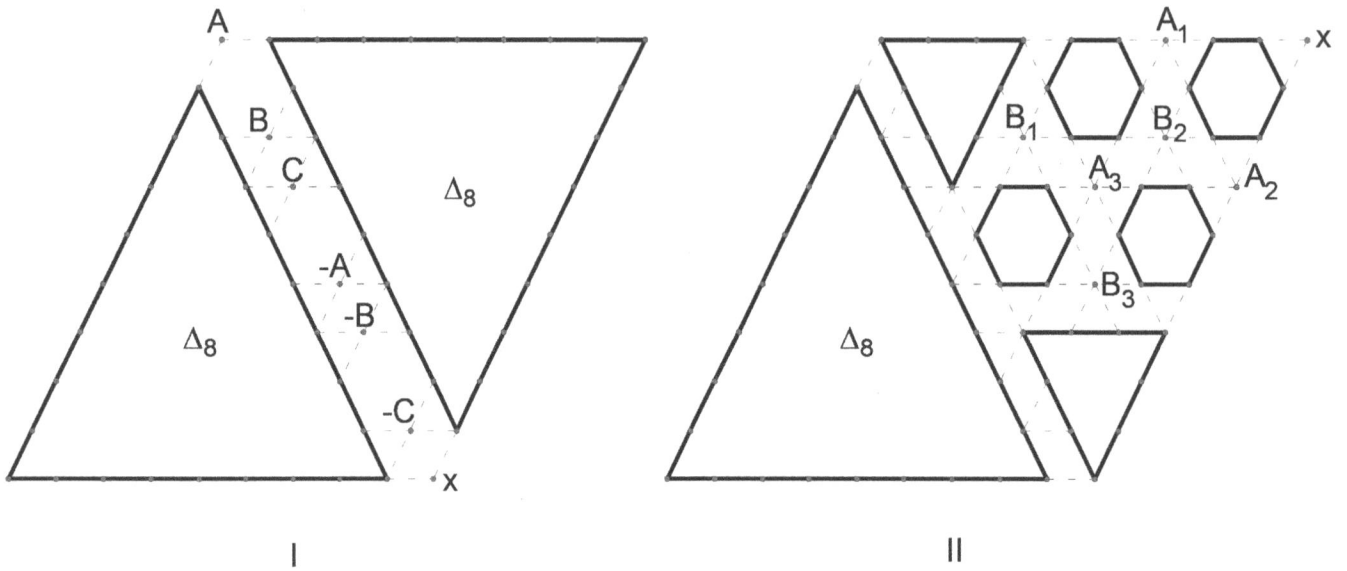

Figure 4.53

In Figure 4.53 the two Types of $\prod_4 R_9$ rhombs are illustrated, each with a simple 0-value dissection which covers all points except the one labeled x. For both versions I and II an H_3 pattern can easily be constructed which assigns the value 1 to point x. As a consequence the value of the entire rhomb will be 1. The conclusion is that neither version of R_9 is nilpotent.

4.9 $\prod_4$ pentagons and hexagons

4.9.1 Regular $\prod_4$ hexagons

In Figure 4.54(I) a simple 0-value dissection shows that H_2 is nilpotent, and in Figure 4.54(II) the only $\prod_4 H_3$ hexagon is illustrated with a locally nilpotent assignment. The value of this assignment is 4, so we can conclude that H_3 is nonnilpotent.

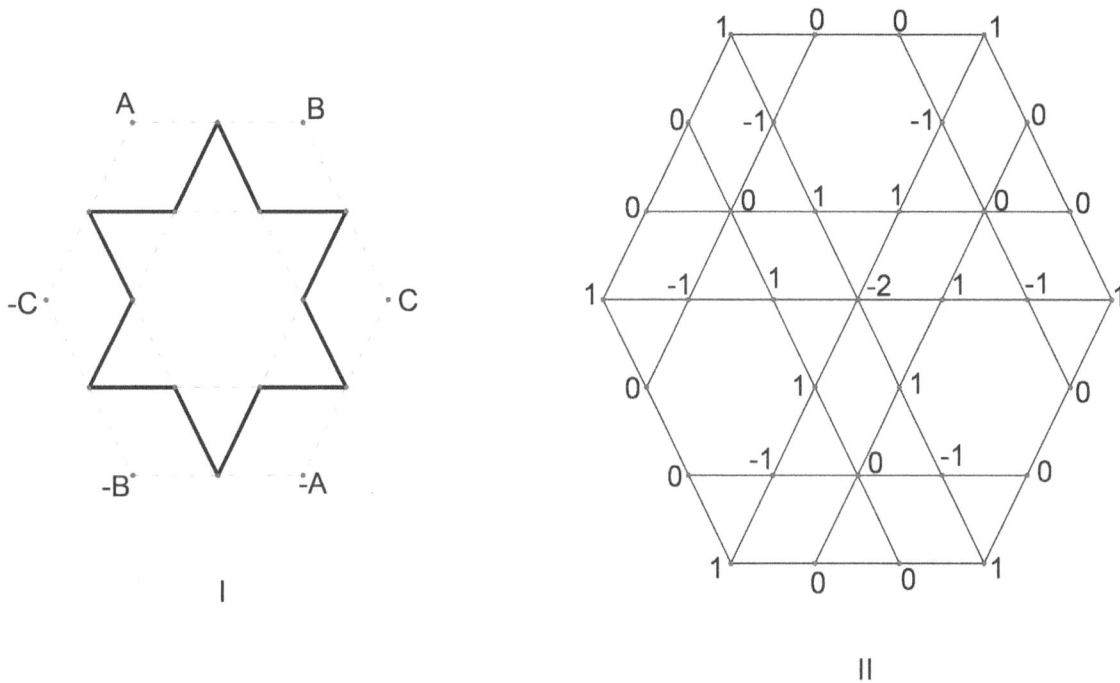

Figure 4.54

In Figure 4.1 we have the construction of a nilpotent H_4. The unique nilpotent H_5 hexagon is illustrated in Figure 4.55 with a simple 0-value dissection. The only H_6 hexagon in the Π_4 plane is illustrated in Figure 4.56. It is equipped with a simple 0-value dissection which however misses the central point x. As in the above argument regarding the R_9 rhomb, we can conclude that H_6 is nonnilpotent.

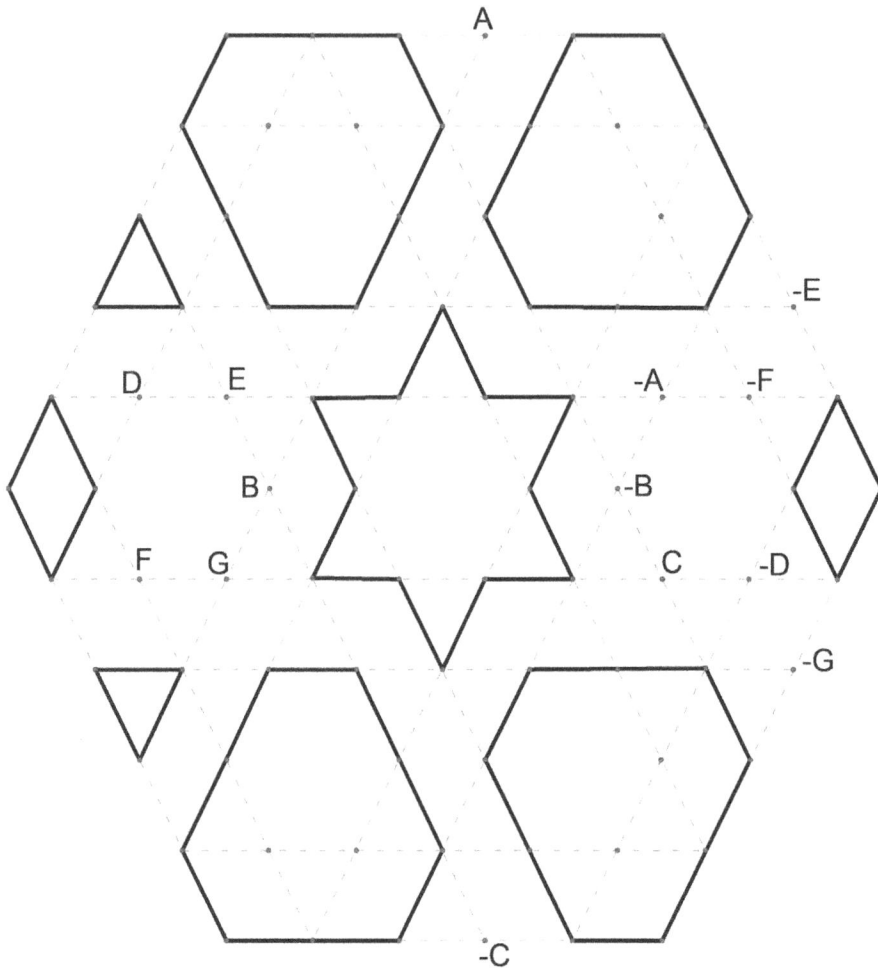

Figure 4.55

In Figure 4.57 we develop a scheme for determining hexagon Types in $\prod_4$. Two locations for positive slope diagonals are indicated by α_1, α_2 and two locations for horizontal diagonals are indicated by β_1, β_2. The structure of a $\prod_4$ hexagon is entirely determined by which of these diagonals are chosen. If $\{\alpha_1, \beta_1\}$ is chosen then a $\prod_4$ hexagon H_m exists iff $m = 2+3t$, $t \geq 0$; if $\{\alpha_2, \beta_2\}$ is chosen then then a $\prod_4$ hexagon H_m exists iff $m = 1+3t$.

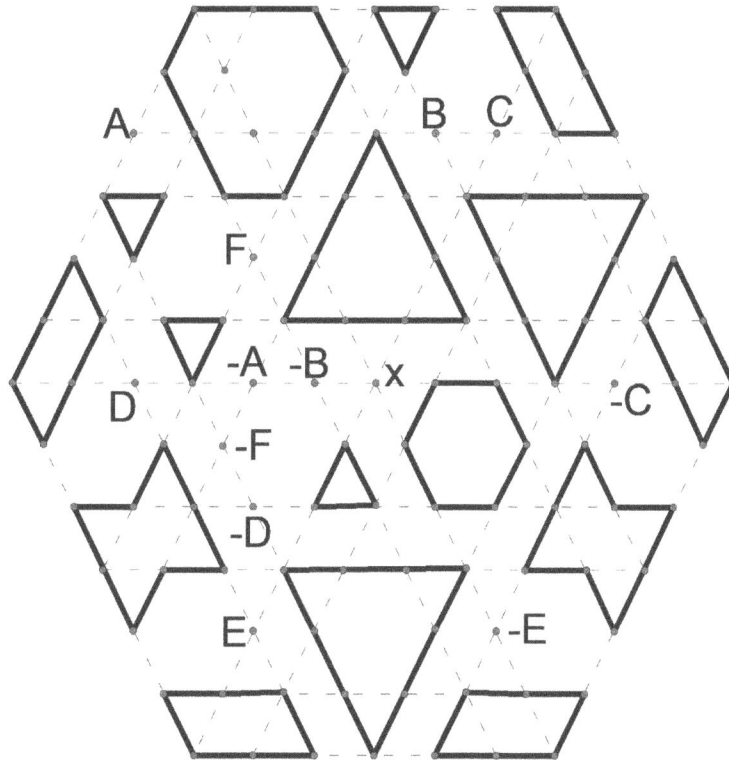

Figure 4.56

Π_4 hexagons exist iff either $\{\alpha_1, \beta_2\}$ or $\{\alpha_2, \beta_1\}$ is chosen, and these two choices result in isomorphic hexagons. Hexagons in the first two categories are distinguished by having a centrally located H_1, we call them *Type(I) central* if $\{\alpha_1, \beta_1\}$ is chosen, and *Type(II) central* if $\{\alpha_2, \beta_2\}$ is chosen. Hexagons in which $\{\alpha_1, \beta_2\}$ or $\{\alpha_2, \beta_1\}$ are chosen occur iff the side has length a multiple of 3. These have a point at center, instead of an H_1, as in Figure 4.56.

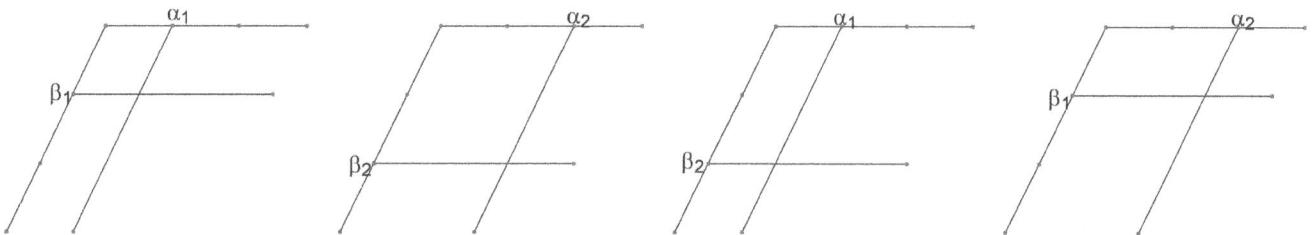

Figure 4.57

In Figure 4.58 a scheme for showing that central hexagons of either Type are nilpotent is illustrated. The scheme involves the extension $H_1 \rightarrow H_2 \rightarrow H_4 \rightarrow H_5 \rightarrow H_7 \rightarrow H_8 \rightarrow H_{10} \rightarrow \dots$. Which covers all Type(I) and Type(II) central hexagons.

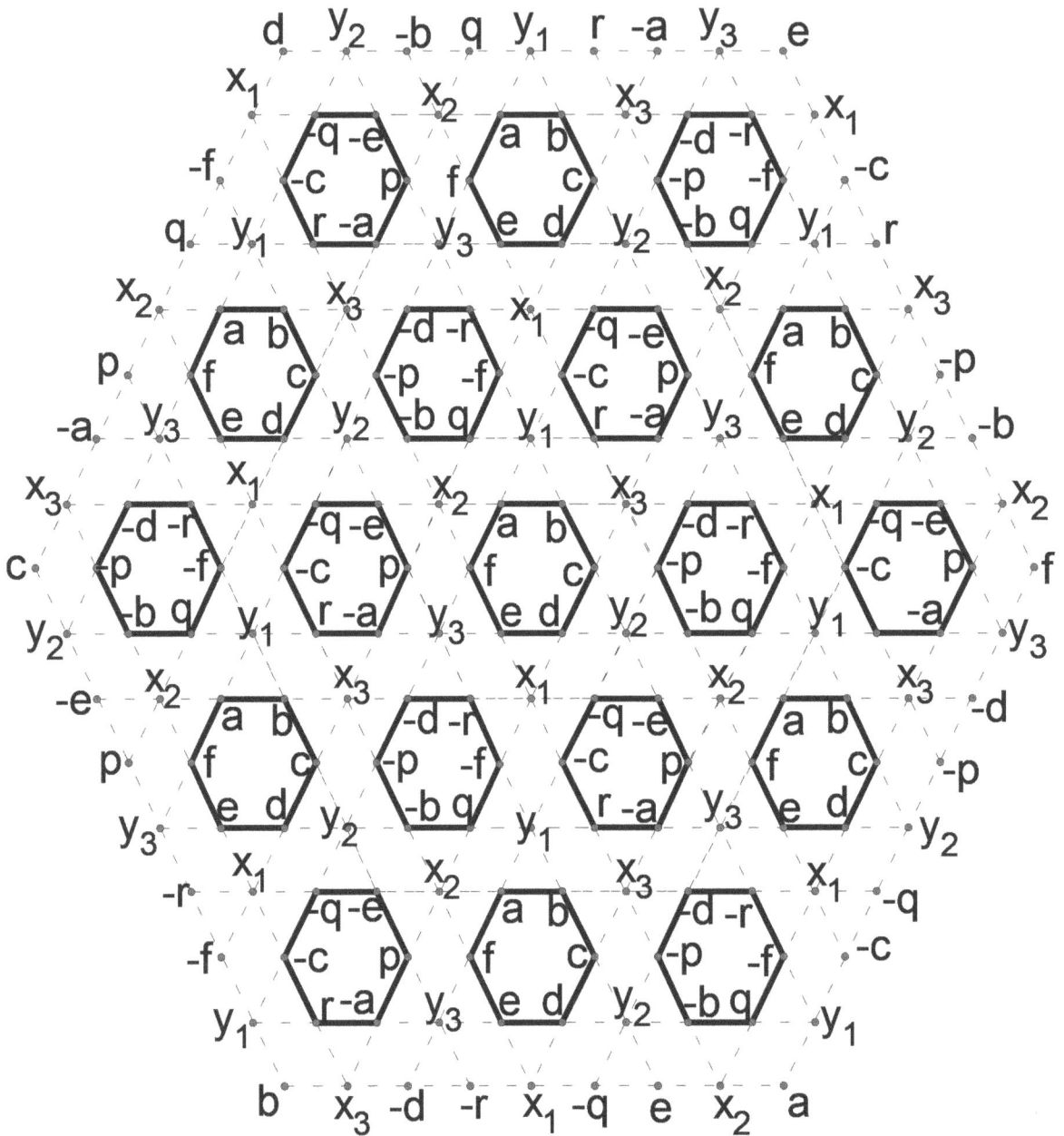

Figure 4.58

In Figure 4.58 it should be observed that the boundary values for H_m, (m= 1, 2, 5, 7, 8), sum to 0. If we skip m edges while traversing the boundary of H_m the sequences of 6 values encountered sum to 0. For example, for the boundary of H_8 the sequences are:

$(y_2, x_1, y_3, x_2, y_1, x_3)$
$(-b, -c, -d, -e, -f, -a)$
$(q, r, -p, -q, -r, p)$
$(y_1, x_3, y_2, x_1, y_3, x_2)$
$(r, -p, -q, -r, p, q)$
$(-a, -b, -c, -d, -e, -f)$
$(y_3, x_2, y_1, x_3, y_2, x_1)$
$(e, f, a, b, c, d),$

all of which clearly sum to 0. The *interboundary points* which lie between the boundary of H_7 and H_8 also form sequences whose total sum is 0:

$(p, f, c, -p)$
$(q, a, d, -q)$
$(r, b, e, -r)$
$(-p, c, f, p)$
$(-q, d, a, q)$
$(-r, e, b, r).$

The same phenomena occur for boundary and interboundary values for every H_m, ($m = 1 + 3t$ or $m = 2 + 3t$, $t \geq 0$, i.e., for every central regular $\prod_4$ hexagon. We thus have the following result.

Theorem 4.20 *Every central regular $\prod_4$ hexagon is nilpotent.*

It remains to consider the noncentral $\prod_4$ hexagons. In Figure 4.59 these hexagons are illustrated in a nesting.

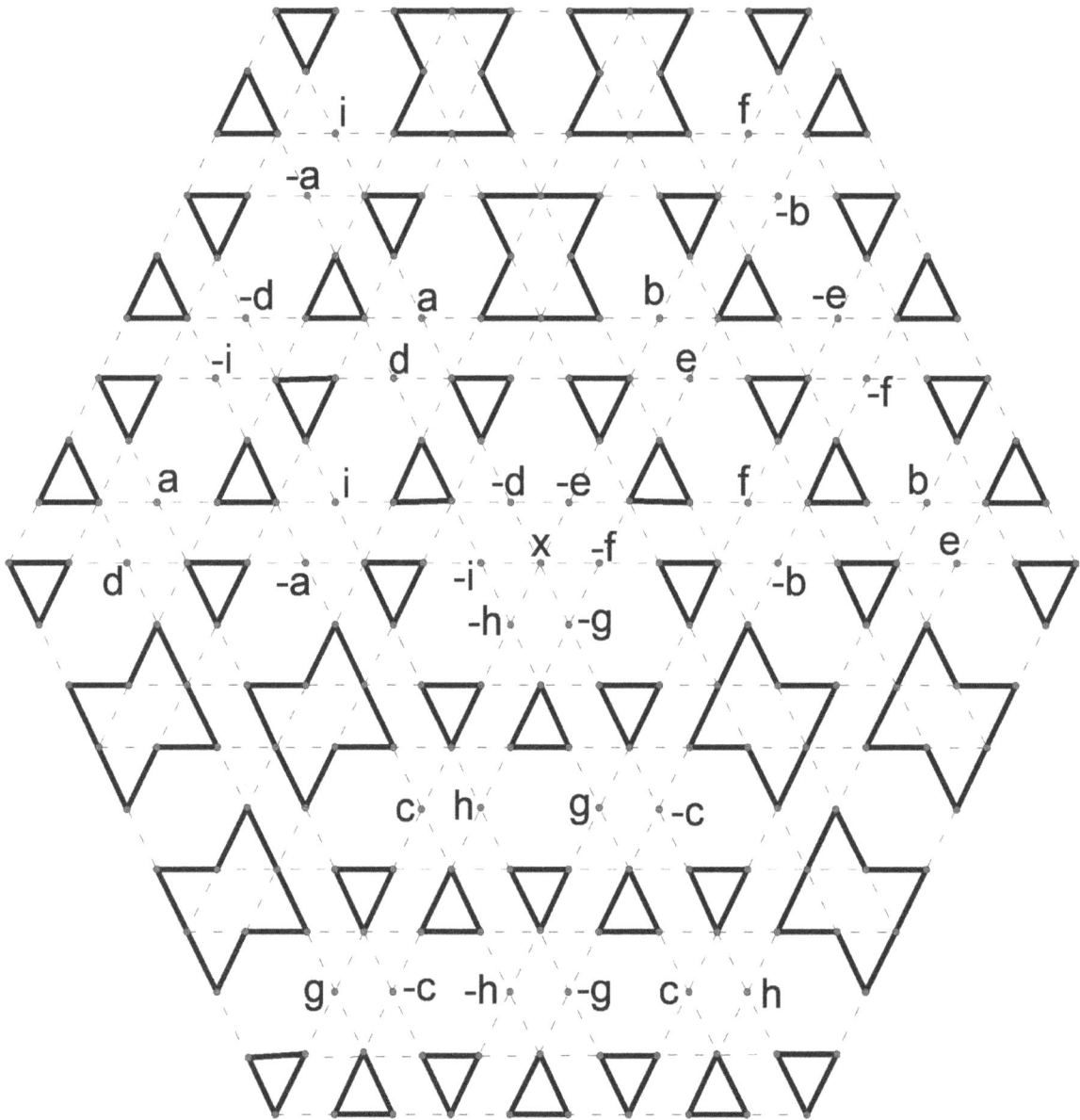

Figure 4.59

The Figure provides a partial simple 0-value cover which misses only the central point x for each of H_6, H_9. Any pattern which assigns a nonzero value to x will show that H_6, H_9 are nonnilpotent. For H_3 and for larger hexagons in the class $\{H_{3t} : t \geq 0\}$ the pattern illustrated in Figure 4.60 can be used to prove nonnilpotence. Thus we have a strengthening of Theorem 4.20:

Theorem 4.21 *A $\prod_4$ hexagon is nilpotent iff it is central.*

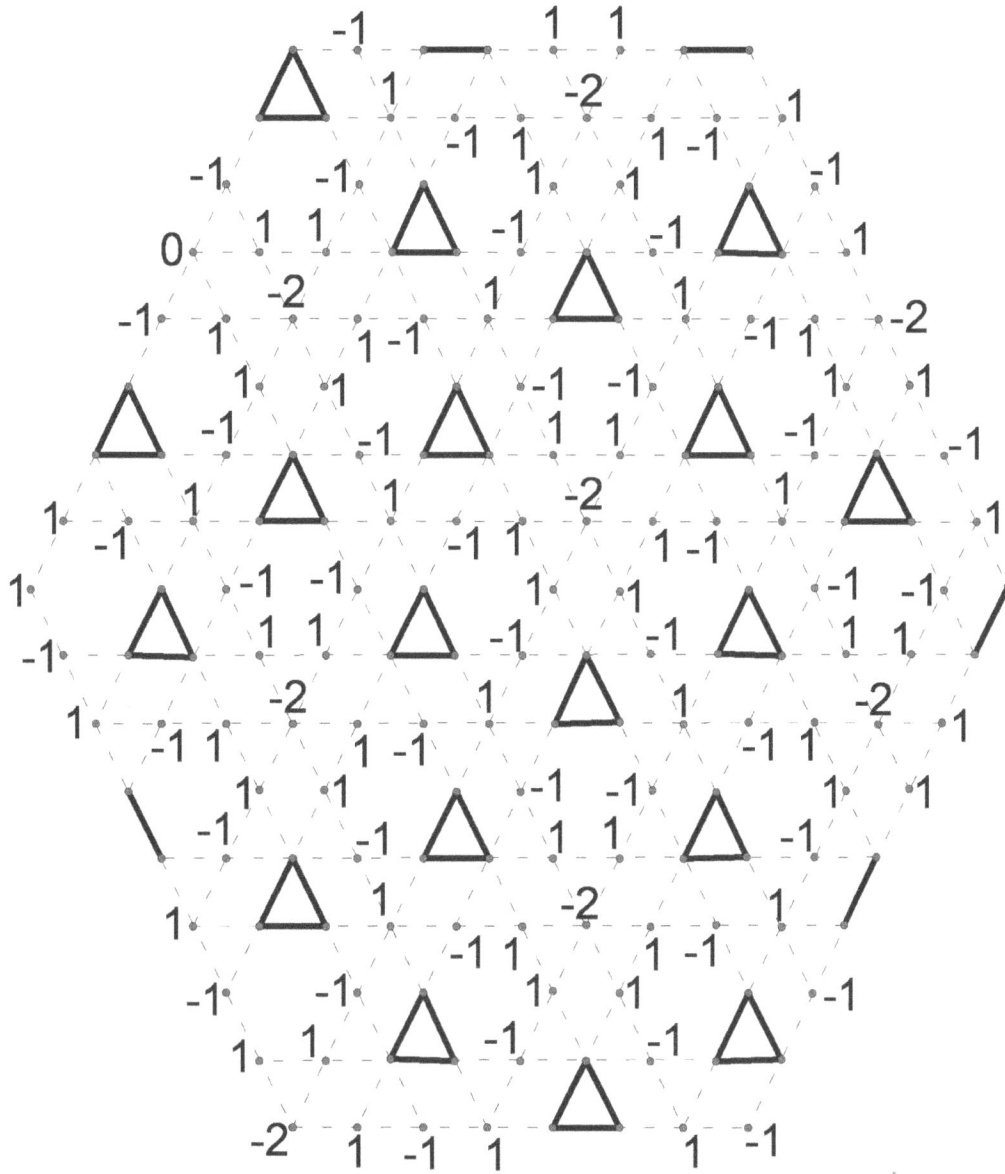

Figure 4.60

4.9.2 δ – polygons and ∈-polygons

In Figure 4.61 a simple 0-value dissection is illustrated. We call this a δ *-dissection* and note that it can be extended to cover the entire $\prod_4$ plane. It hosts a number of nilpotent polygons which we call δ – *polygons*. Among these, we leave it to the reader to find the following Types.

Semiregular δ – polygons. These are nilpotent δ- hexagons in the form H(a,b,c,a,b,c). Examples include H(2,2,6,2,2,6) and extensions such as H(5,2,6,5,2,6).

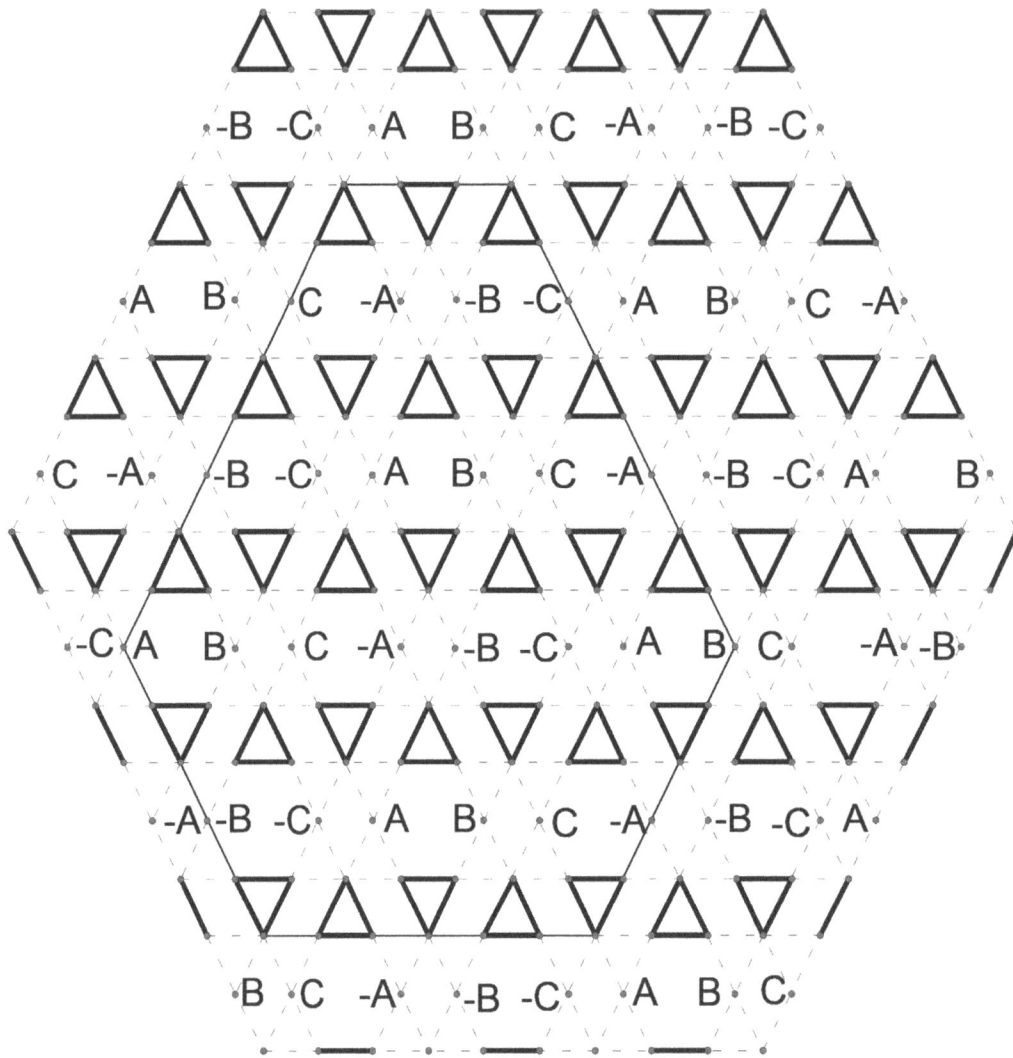

Figure 4.61

(i) *Isosceles δ- hexagons.* These are hexagons formed by two isosceles trapezoids which have a common base. Examples include H(2,8,3,8,2,9) and its extension H(5,8,3,8,5,6). Also H(2,8,6,8,2,12) and its extension H(3,11,8,6,8,11).

(ii) Besides the above hexagons, various δ-trapezoids and δ-triangles can be found, for example: Δ_7, Δ_{10}, Δ_{16} and T(6,4,10), T(6,10,16), T(3,10,13).

(iii) *Isosceles δ-pentagons.* These are pentagons of the form P(a,0,a,b,c,b) where b+c = a, and where the 0 indicates the location of an interior angle of 60°. (We regard a pentagon as a degenerate hexagon in which one side has collapsed to 0 length.) Examples include P(5,0,5,2,3,2) and its extension P(8,0,8,2,6,2).

(iv) *δ-parallelograms.* For example, P(1,2) and its extension P(10,2). Also P(1,5) and its extension P(10,5).

The full determination of δ-polygons we leave as an open question. In a second class of Π_4 polygons all the H_1 hexagons are covered in a dissection and 0-sum point triples which occur at corners of triangles are used to complete the dissection. Point couples and Δ_1 triangles are not allowed. We call nilpotent polygons in this class *ε-polygons* and will now focus our attention on the construction of ε-hexagons.

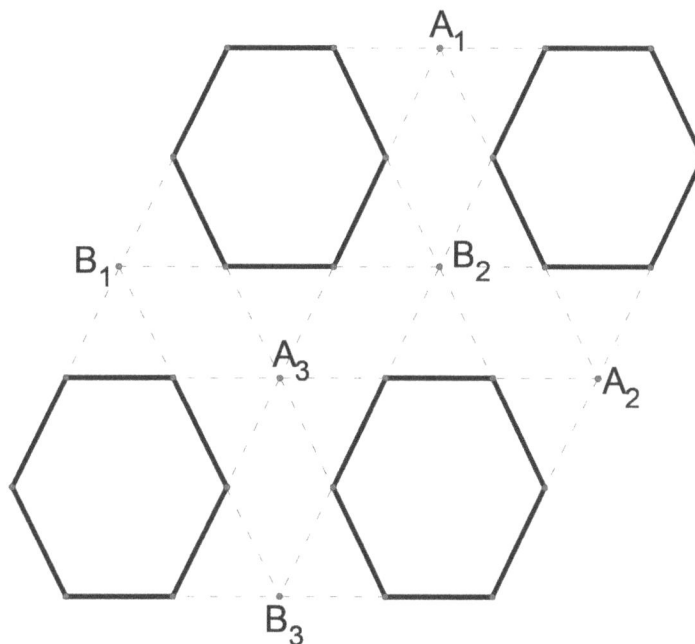

Figure 4.62

If the H_4 in Figure 4.51 is modified slightly by replacing the central star by a hexagon and two point triples at the corners of Δ_3 triangles, we see that H_4 is an ε-hexagon. The nilpotent triangle Δ_3 is also an ε-hexagon along with the semiregular H(1,4,4,1,4,4) illustrated with an ε-dissection in Figure 4.62.

These elements, the Δ_3, H_4, H(1,4,4,1,4,4) are *ε-primes* in the sense that they cannot be dissected into smaller ε-polygons. (Note that although Δ_3 is composite when considered in the general class of nilpotent Π_4 polygons, it is nevertheless ε-prime.) In Figure 4.63 we combine these elements to create an example which contains several *ε-composites.*

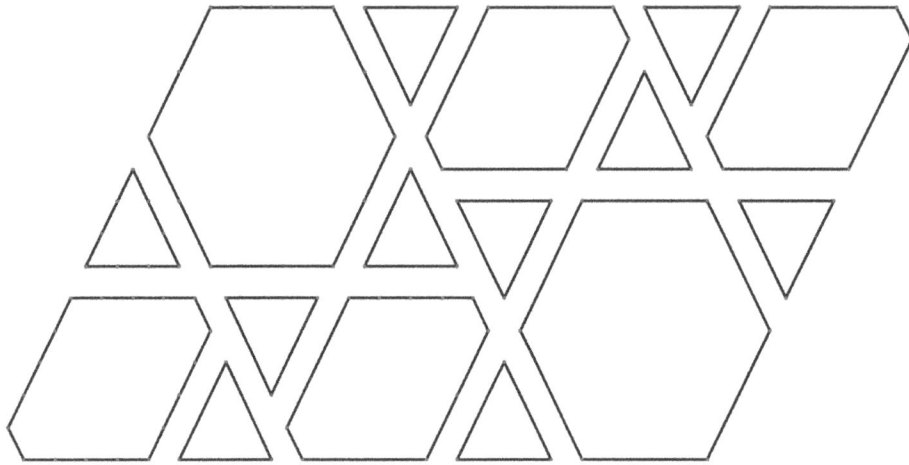

Figure 4.63

In Figure 4.64 an ε-dissection of H_{10} is illustrated. It features the new ε-prime $H(1,10,10,1,10,10)$ as a component. If each $H(1,4,4,1,4,4)$ in this component is replaced by an $H(1,10,10,1,10,10)$ hexagon then the ε-prime $H(1,22,22,1,22,22)$ hexagon is created. In this hexagon the 6 missing points form the corners of two Δ_{12} triangles, and these corners are easily seen to have 0-sum. This enlargement process can be repeated to obtain prime ε-hexagons of arbitrarily large size.

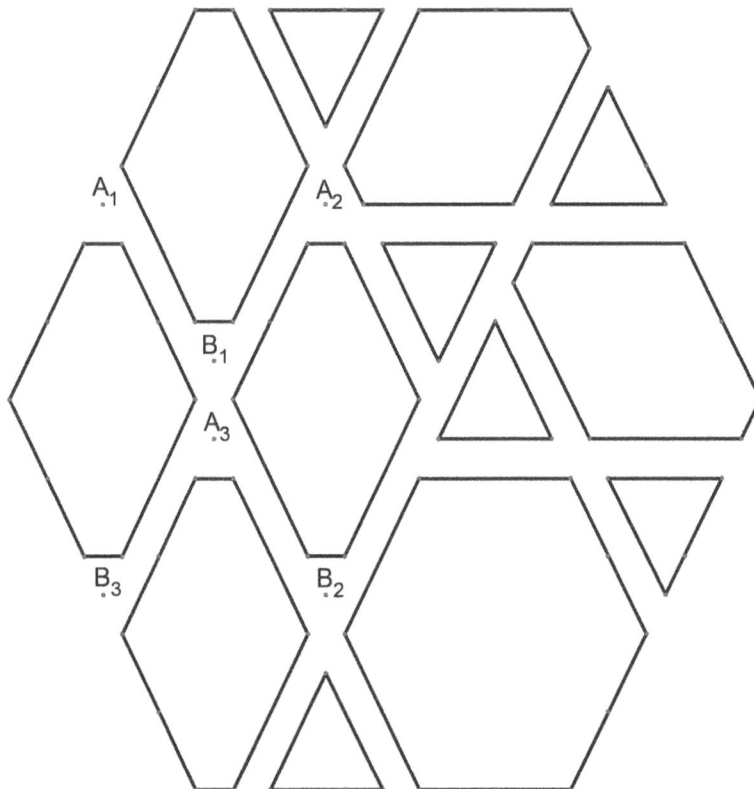

Figure 4.64

It should also be pointed out that the enlargement process, mentioned in Section 4.4, which replaces a Δ_1 triangle and an H_1 hexagon in a $\prod_1$ polygon by a Δ_3 and an H_4 respectively will result in an ε-$\prod_4$ polygon. In this way, for example, the ε-hexagon H_{13} can be created. We conclude this Section with two Questions which merit further research.

Question 4.22 *For which integers m does a regular ε-hexagon H_m exist ?*

Question 4.23 *Besides Δ_3, H(1,4,4,1,4,4), H_4, H(1,10,10,1,10,10) and its enlargements, what ε-primes can be constructed ?*

4.9.3 Other $\prod_4$ hexagons

A simple $\prod_4$ hexagon which is neither a δ nor an ε-hexagon is the H(1,2,3,2,1,4) isosceles hexagon which can be found in three places in both the isosceles hexagon H(1,11,3,11,1,13) in Figure 4.65(I) and the isosceles hexagon H(2,12,2,8,6,8)) in Figure 4.65(II).

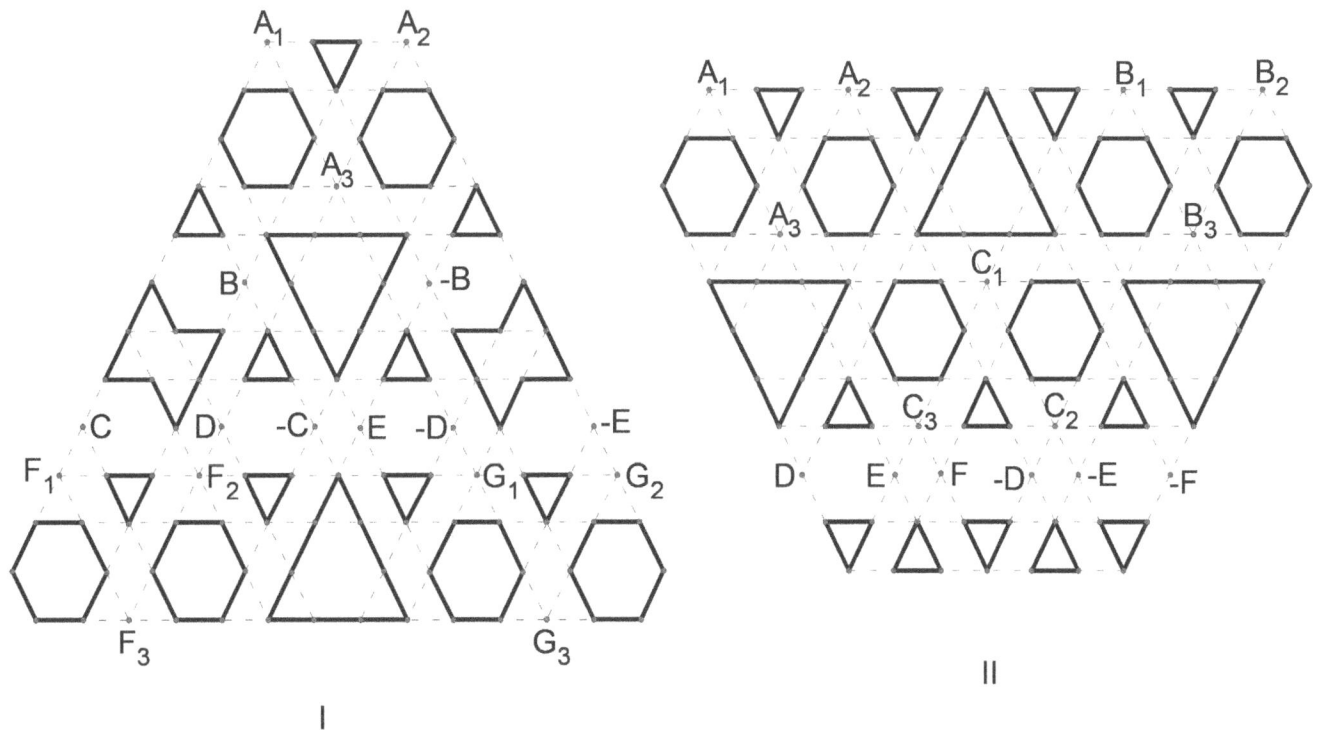

Figure 4.65

Both hexagons in Figure 4.65 represent extensions of a nilpotent isosceles hexagon H(1,2,12,2,1,13). These hexagons can be combined with one on top of the other, to obtain a nilpotent isosceles hexagon H(8,11,3,11,8,6). The H(2,12,2,8,6,8) in Figure 4.65(II) is an extension of the nilpotent isosceles hexagon H(2,12,2,5,9,5).

Two more nilpotent hexagons are illustrated in Figure 4.66. Both are irregular: 4.66(I) is H(3,6,11,7,2,15) and 4.66(I) is H(6,2,13,2,6,9). Both feature multiple occurences of H(1,2,3,2,1,4) components. Also, both are extensions of a nilpotent irregular H(3,6,2,7,2,6). In 4.66(I) a nilpotent irregular H(5,4,4,7,2,6) can be found.

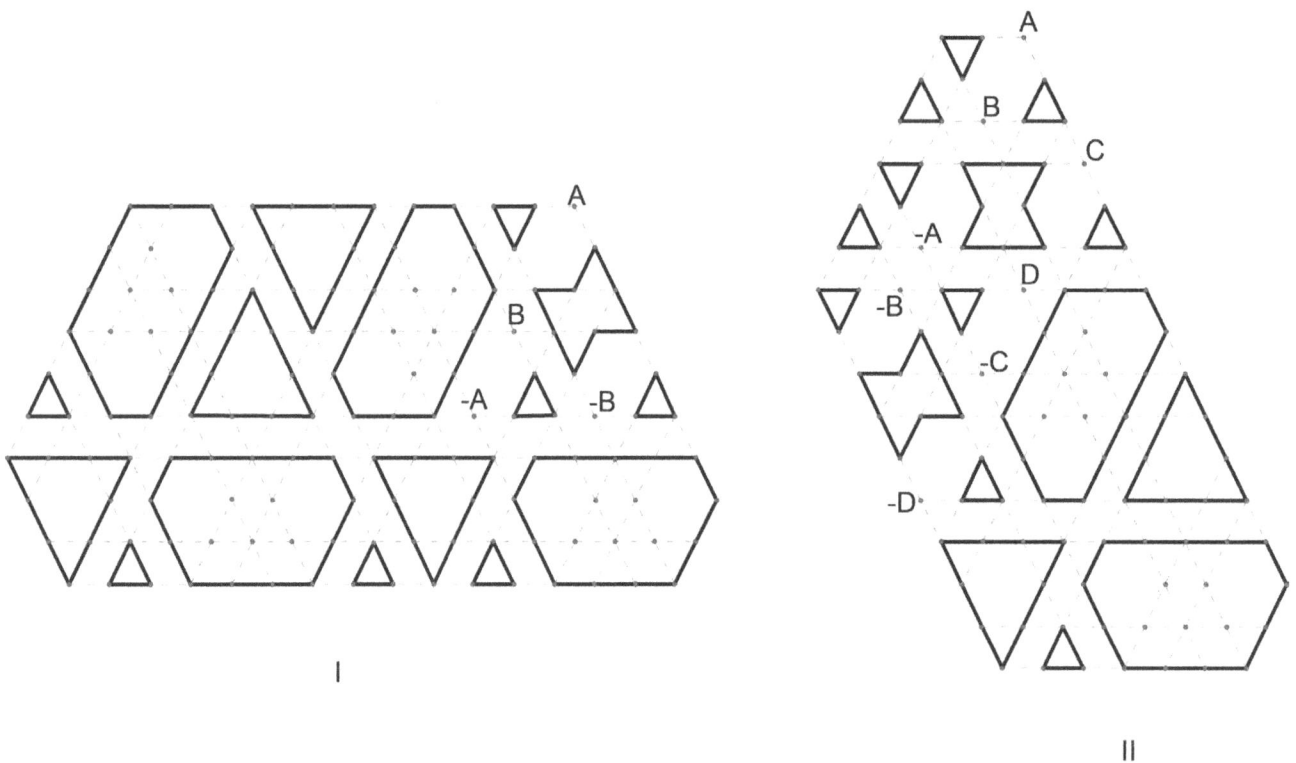

Figure 4.66

In Figure 4.67(I) yet another nilpotent hexagon H(7,8,3,8,7,4) is illustrated which contains several H(2,3,2,1,4,1) components. It is an extension of an H(2,8,3,8,2,9).

Figure 4.67(II) features a nilpotent semiregular Π_4 hexagon H(3,4,4,3,4,4) with two prime H(1,2,1,2,1,2) components. It is an extension of the isosceles H(2,5,2,4,3,4). In Figure 4.68 a nilpotent isosceles H(2,11,2,11,2,11) which contains both H(1,2,1,2,1,2) and H(2,3,2,1,4,1) components.

In our study of nilpotent $\prod_4$ polygons we have restricted our attention primarily to convex polygons. Among these only triangles and regular hexagons have been fully determined. The following categories of nilpotent hexagons we leave open for further research.

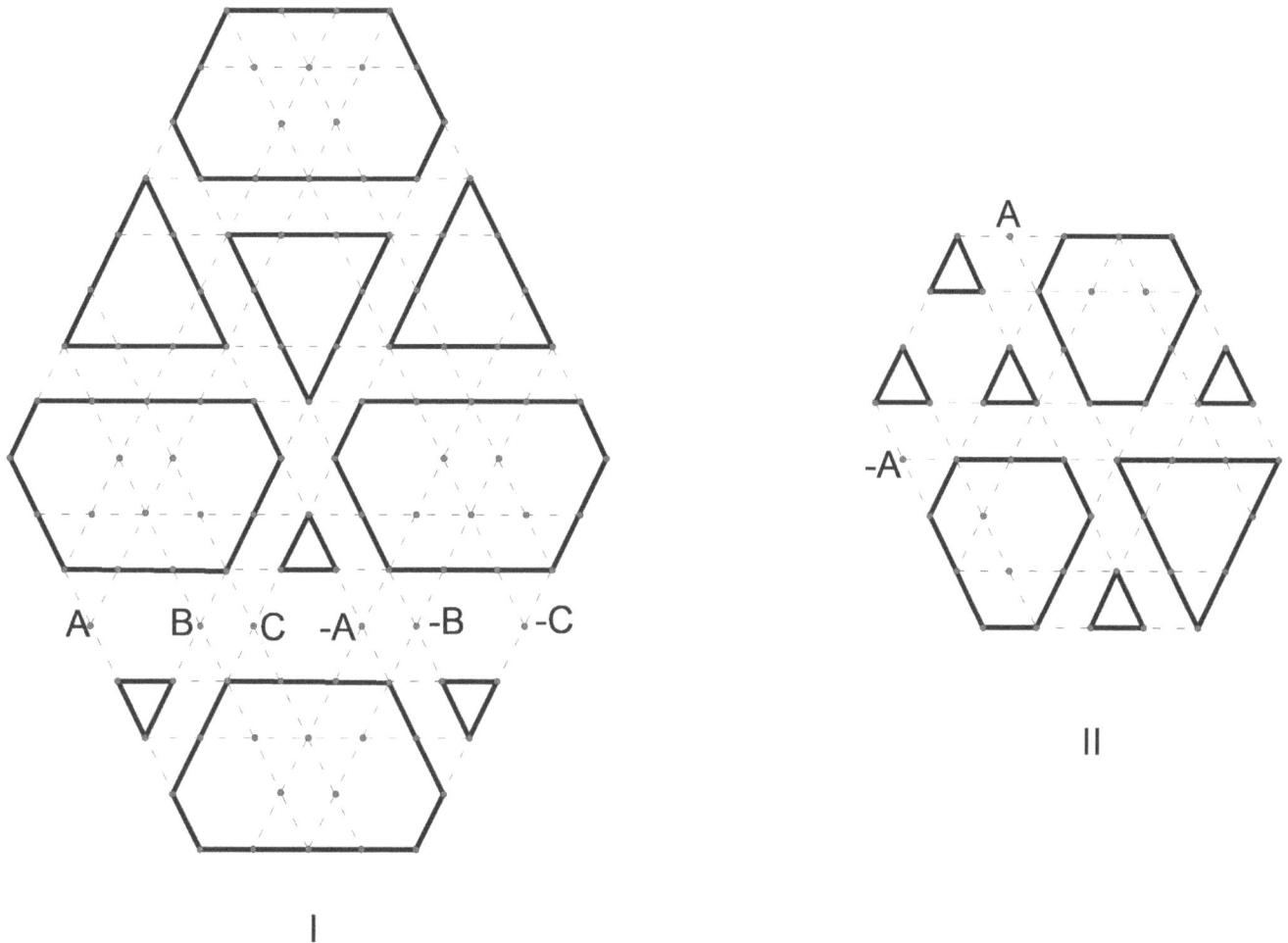

Figure 4.67

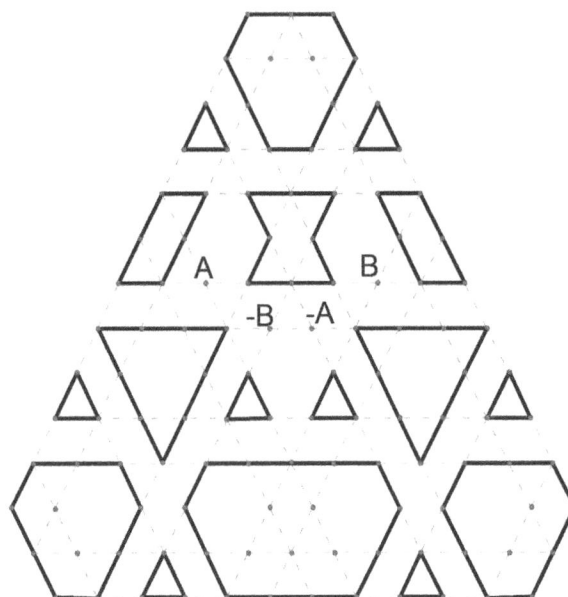

Figure 4.68

Near-regular hexagons. These are semiregular with two pairs of opposite and equal sides. H(1,4,4,1,4,4), Figure 4.62, and H(2,2,6,2,2,6), see Figure 4.61, are examples.

(i) *Alternating hexagons.* These are $\prod_4$ hexagons with the form H(a,b,a,b,a,b). H(1,2,1,2,1,2), and H(1,10,1,10,1,10), Figure 4,64, and H(2,11,2,11,2,11), Figure 4.68, are examples of nilpotent alternating hexagons.

Isosceles hexagons. These have the form H(a,b,a,c,d,c) with a+b = c+d. We have seen many examples of these.

4.10 The $\prod_4$ game

The playing board for the $\prod_4$ game is the $\prod_4$ hexagon H_7. This hexagon has 150 points and is provided with 150 discs in 15 groups numbered 0, ..., 9. For scoring configurations we think it is best to restrict to nilpotent triangles and hexagons. Also the wedge W_1 and the star J_1 should be included. Commonly occurring scoring configurations include:

$$R_1, H_1, \Delta_1, \Delta_3, H(1,2,1,2,1,2), H(2,3,2,1,4,1), H_2, W_1, J_1.$$

These are all intrinsically nilpotent. Larger intrinsically nilpotent structures can occur but are not likely since, ordinarily, the accumulation of dead points prevents their construction. The following conditionally nilpotent polygons may also be allowed as scoring configurations:

$$\Delta_2^*, \Delta_3^*, \Delta_4^*, H^*(1,1,4,1,1,4), H^*(2,2,3,2,2,3).$$

We have indicated the conditional status of these configurations with an (*). Finally it is interesting to include scoring configurations which we call *wedge complexes*. Several examples of these are illustrated in Figure 4.69. These do occur commonly in the play of the game and can be denoted by W_m where m denotes the number of W_1 wedges which constitute the complex.

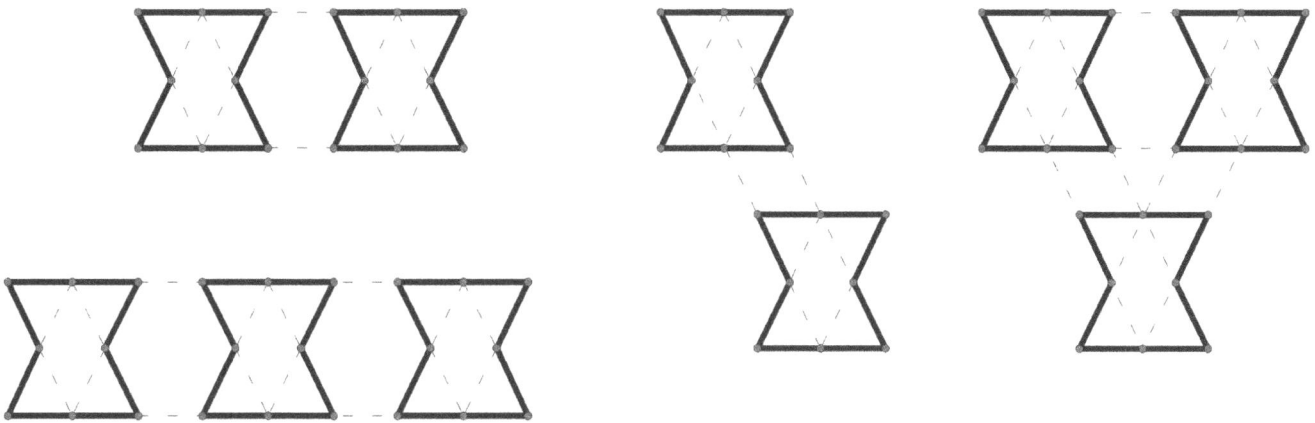

Figure 4.69

The wedge complexes in Figure 4.69 are all composite. In Figure 4.70(I) a prime wedge complex is illustrated along with a 0-value dissection. It consists of a chain of three overlapping wedges. Another system of three overlapping wedges is illustrated in Figure 4.70(II). This polygon is conditionally nilpotent, with A+B = 0 the requirement for nilpotence.

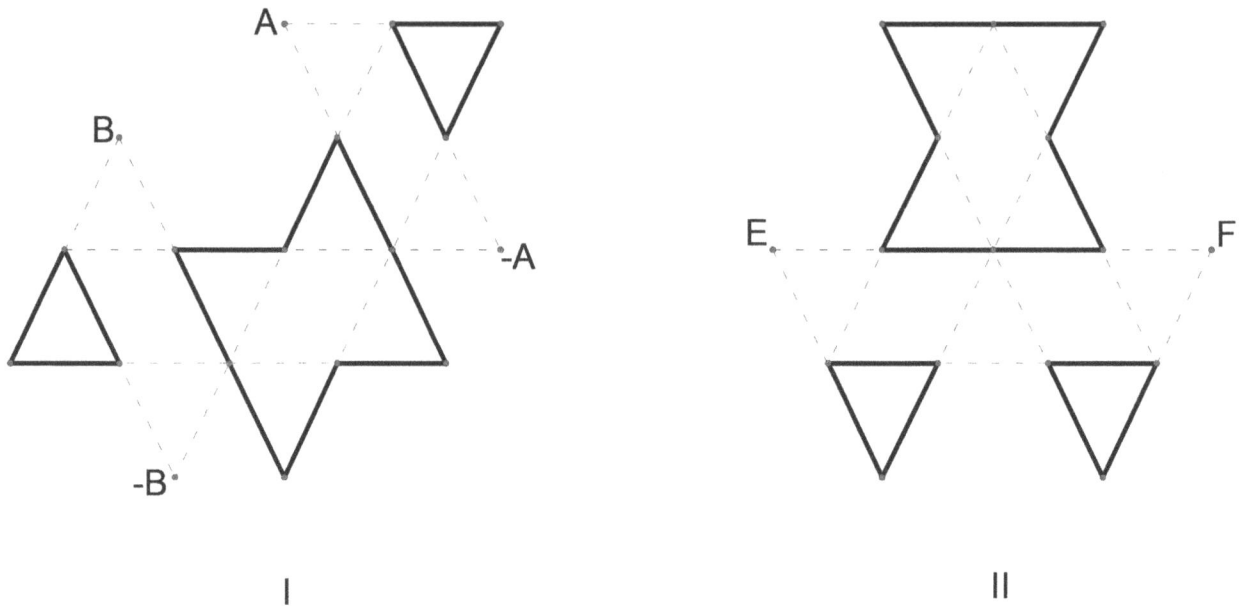

Figure 4.70

As usual nilpotent configurations over Z are imitated over Z_5 in the game, and thus the total value of the discs in a scoring configuration must be a multiple of 5. This multiple of 5 constitutes the score earned when a scoring configuration is completed. Also, as usual, only a disc which completes a multiple of 5 can be placed in a potential scoring position.

A *dead point* occurs whenever the play of a disc on a single point completes two scoring configurations with conflicting requirements. In case the requirements are nonconflicting then a *double score* occurs. For a double score to occur, it is also necessary that the scoring configurations are not one contained in the other. If one is contained in the other, then only the larger structure contributes a score.

A commonly occurring double score, as well as a dead point are illustrated in Figure 4.71. The point y is a dead point since a 1 or a 6 is required to complete the Δ_1 triangle, but a 2 or a 7 is required to complete the rhomb R_1. If a 1 or a 6 is played at point x, then a double score occurs: two $H(1,2,1,2,1,2)$ hexagons as well as an $H(1,4,1,2,3,2)$ are completed. One of the $H(1,2,1,2,1,2)$ hexagons is contained in the $H(1,4,1,2,3,2)$, so the total score is obtained by adding the score (90 or 95) for the $H(1,4,1,2,3,2)$ and the score (45 or 50) for the upper $H(1,2,1,2,1,2)$.

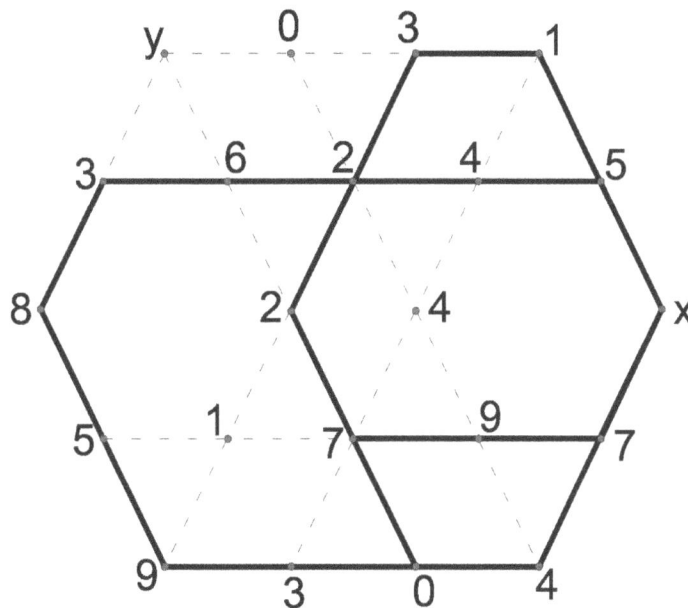

Figure 4.71

We will describe now the play for two players. The game works fine also with 3 or 4 players with the same rules. Each player draws 10 discs from the boneyard, and an additional disc is drawn and played on one of the points of the central H_1. Whenever a disc is played it must be played adjacent to an already played disc. Two types of play are possible: scoring and nonscoring.

When a nonscoring disc is played the players turn ends and he/she draws from the boneyard to obtain a full complement of 10 discs. On the other hand, scoring discs can be repeatedly played in the same turn. Unless a player uses all 10 discs as scoring discs, *the final play for the turn must be a nonscoring disc.*

Players alternate turns until no more scoring discs can be played. In case one player plays all his/her discs and the boneyard is empty then he/she must pass, and the other player can play repeatedly until no more plays are possible.

Recall that a *blocking play* occurs when a nonscoring disc creates an adjacent dead point when placed. Sufficient dead points occur as a result of the placement of scoring discs, so we feel that the game is improved by disallowing blocking. Playing a nonscoring disc in a region occupied by a dead point should be disallowed.

CHAPTER 5

The Π_5 Game

5.1 Scoring configurations

The game for this Chapter will be played on the section of the Π_5 *plane* illustrated below.

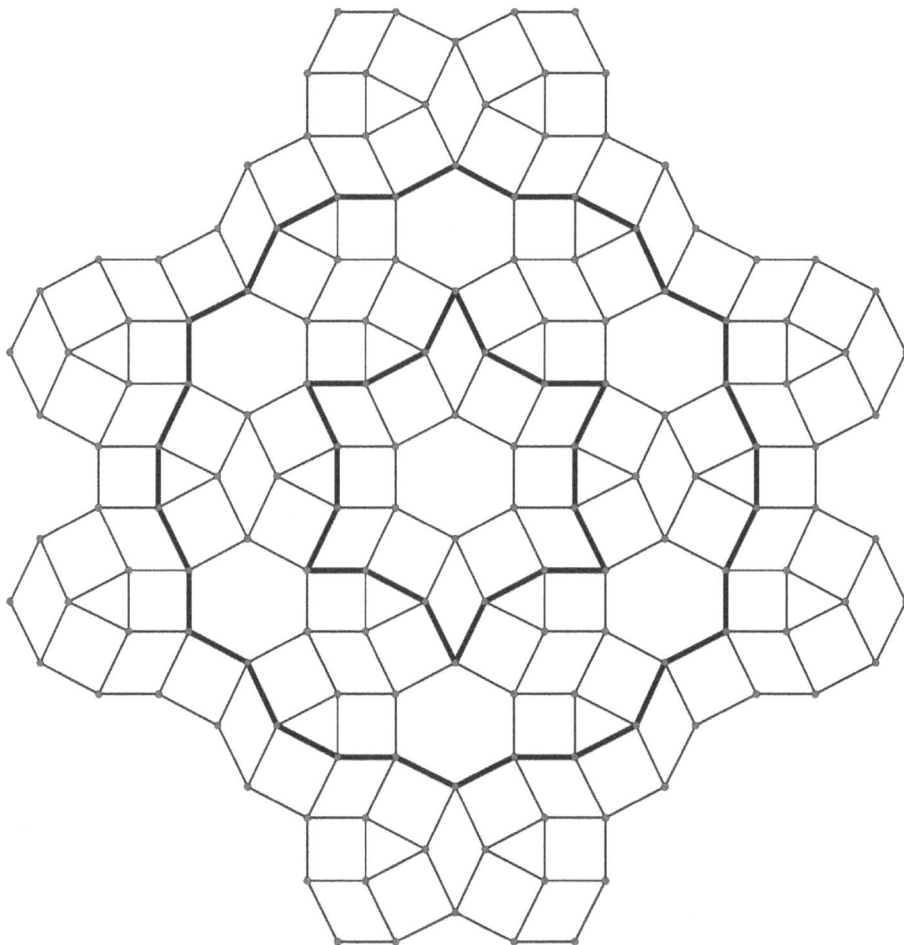

Figure 5.1

The inner circle of bold lines constitutes a barrier and the outer circle of bold lines indicates a possible boundary. Extensions outside this boundary indicate a possibility for a larger game. The tiles H, (hexagon), Δ, (triangle), R, (rhomb), and M, (square), constitute the basic scoring configurations. In Figure 5.2 some of the smaller scoring configurations are illustrated.

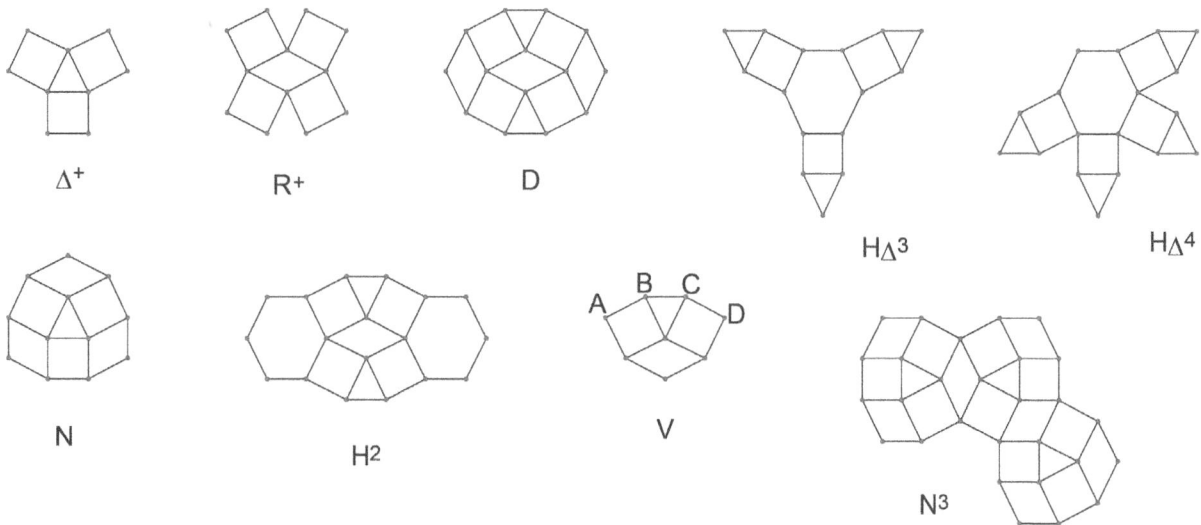

Figure 5.2

In this Figure, Δ⁺ and R⁺ are prime and the 7-gon V is conditional, (A+B+C+D must be 0.) The remaining configurations are composite. $H\Delta^3$ and $H\Delta^4$ are hexagons with 3 and 4 *arms* respectively. A hexagon may have any number from 1 to 6 arms. We think that all should be scoring configurations except for the hexagon with only one arm. The N^3 in Figure 5.2 contains two N^2 configurations. Configurations which consist of 4 or 5 N polygons, denoted N^4 and N^5 respectively can also occur. As usual, all scoring configurations correspond to nilpotent or conditionally nilpotent polygons.

A *stream* is a sequence of alternating squares and rhombs. Streams which stretch from one boundary edge to another should be considered as scoring configurations. In the small game there are 6 such streams each containing 9 tiles. These we denote by L_9. In the large game streams L_7, L_{11}, L_{13} and L_{15} can be found, all of which we feel should be considered as scoring configurations. In Figure 5.3 some larger scoring configurations are illustrated.

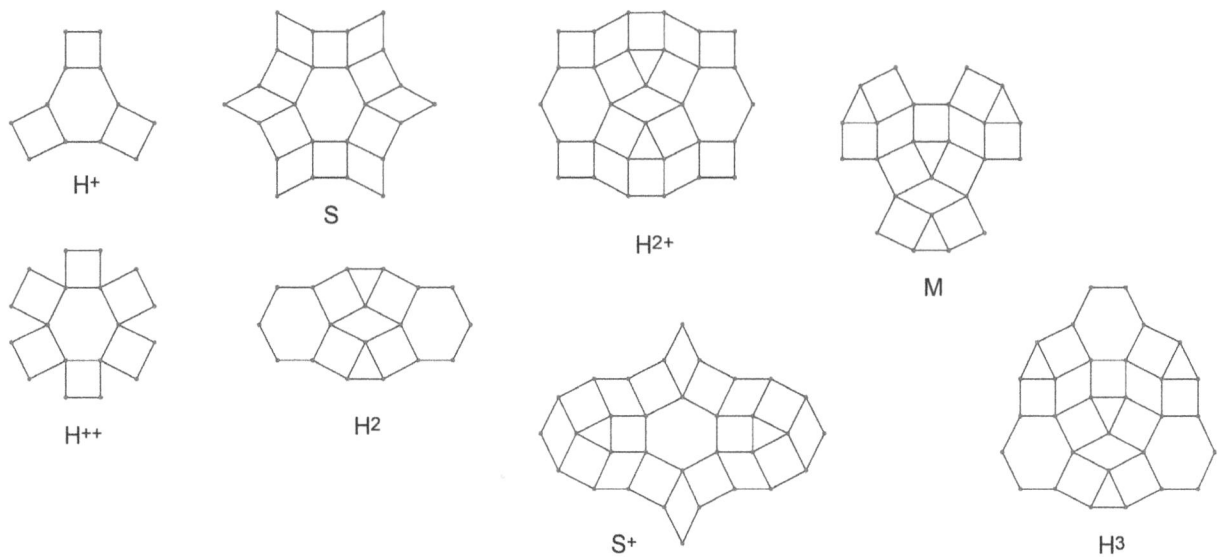

Figure 5.3

Figures 5.2 and 5.3 illustrate just some of the possible scoring configurations. For scoring to be less cumbersome we give preference to configurations which exhibit some kind of symmetry. The larger structures like H^{2+}, M, H^3, S, S^+ are more likely to occur in the large game. H^{++}, H^{2+} and M are examples of prime configurations, the other polygons in Figure 5.3 are composite.

The play of a nonscoring disc which creates an adjacent dead point is called a *blocking play*. Placing a blocking disc is an important strategy which limits the scoring possibilities for the opponent. It is possible to prohibit blocking plays and thus increase the possibility of obtaining larger scoring configurations. This works fairly well in the smaller game, but for the large game the creation of very large scoring configurations would slow the game considerably just due to the time required to tabulate scores. Thus for the larger game we think that blocking plays should be allowed. Playing a nonscoring disc in a region occupied by a dead point should be disallowed.

5.2 The play

The play is similar to our other games but for the readers convenience we review it here. At the outset it is important to specify exactly which configurations are scoring. The following is a simple solution: except for streams, hexagons with 2 or more arms, and bunches of 9-gons, for example N^3 as in Figure 5.2, allow as scoring configurations only nilpotent polygons which have either an axial or a rotational symmetry. For 9-gon bunches it may be better to require that any two adjacent 9-gons have a rhomb in common. It should also be specified at the outset whether blocking is allowed or not.

The materials for the game include a playing board as in Figure 5.1, and 150 discs with values {0, ..., 9) in 15 groups. (For the small game only 80 discs are required.) To start the play each player draws 10 discs from the boneyard, and then another disc is drawn from the boneyard and placed on a point of the central hexagon.

As usual, players alternate turns and must play their discs adjacent to an already played disc. In the same turn a sequence of scoring discs can be played followed by a nonscoring disc. The turn ends either after the nonscoring disc is played or all 10 discs in the hand have been played. Only when the turn ends is the player allowed to draw from the boneyard to replenish his/her hand to a full complement of 10 discs. All points in and on the boundary of the central star barrier must be covered by a disc or by a dead point marker before any disc can be played outside the barrier.

A *scoring position* is a point, which if covered by an appropriate disc, completes a scoring configuration. A scoring position can *only be occupied by* a disc which creates a scoring configuration. As usual, to be a scoring configuration it is necessary that the constituent discs have values which sum to 0 (mod 5), i.e., sum to a multiple of 5. That multiple constitutes the score for the play.

A *double score* occurs when two scoring configurations, neither of which is contained in the other, are completed by the play of a disc. In this case the scores for each configuration are added to obtain the final score for the play. If one of the configurations is contained in the other, then only the score for the larger configuration is tabulated.

Players alternate turns until the boneyard is exhausted and neither player has a legal move. If a player has no legal move then he/she must pass and the other player may continue to play.

5.3 Properties of the $\prod_5$ plane

It is easily seen that the dimension of the $\prod_5$ plane is infinite. Even the small game requires 16 generators. Recall that a locally nilpotent assignment is *free over a given plane* in case such an assignment over Z can be made without any obstructions. (An obstruction occurs when the same point requires two different values.) In constructing a locally nilpotent assignment it is important to cover any points whose value is forced before assigning values to any other point. Using this approach it is seen that $\prod_5$ has a free locally nilpotent assignment.

CHAPTER 6

The Π_6 game

6.1 The Π_6 plane

A section of the Π_6 plane is illustrated in Figure 6.1.

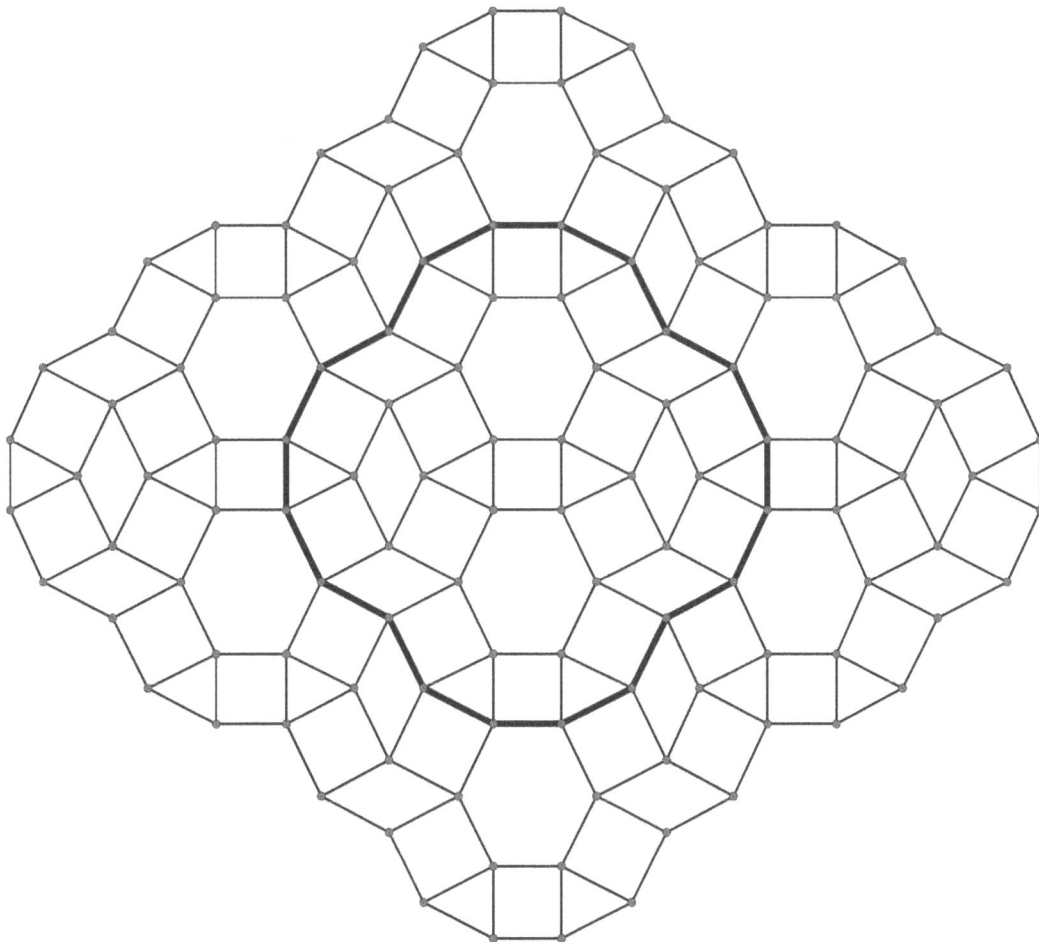

Figure 6.1

Π_6 does have a free locally nilpotent assignment. It is however dependent on a 0-sum point couple. Consider just the central configuration in Figure 6.2. The tetradecagongon, which we denote by T, is indicated in bold in Figure 6.2. It has a vertex cover consisting of two rhombs and four triangles, so it is nilpotent. The hexagon surrounded by 6 squares, which is contained in T, is also nilpotent. As a consequence the opposite boundary points a, h of T sum to 0. A locally nilpotent assignment for T is easily obtained. The values p, q can be chosen freely, and in the end, the values s, t must satisfy (1) s+t = p+q, (from the square), and (2) s+t = a+p+h+q (from the hexagon). Since a+h = 0, these equations are both satisfied. It follows that the central configuration, (C) has a free locally nilpotent assignment, and indeed so does the entire Π_6 plane.

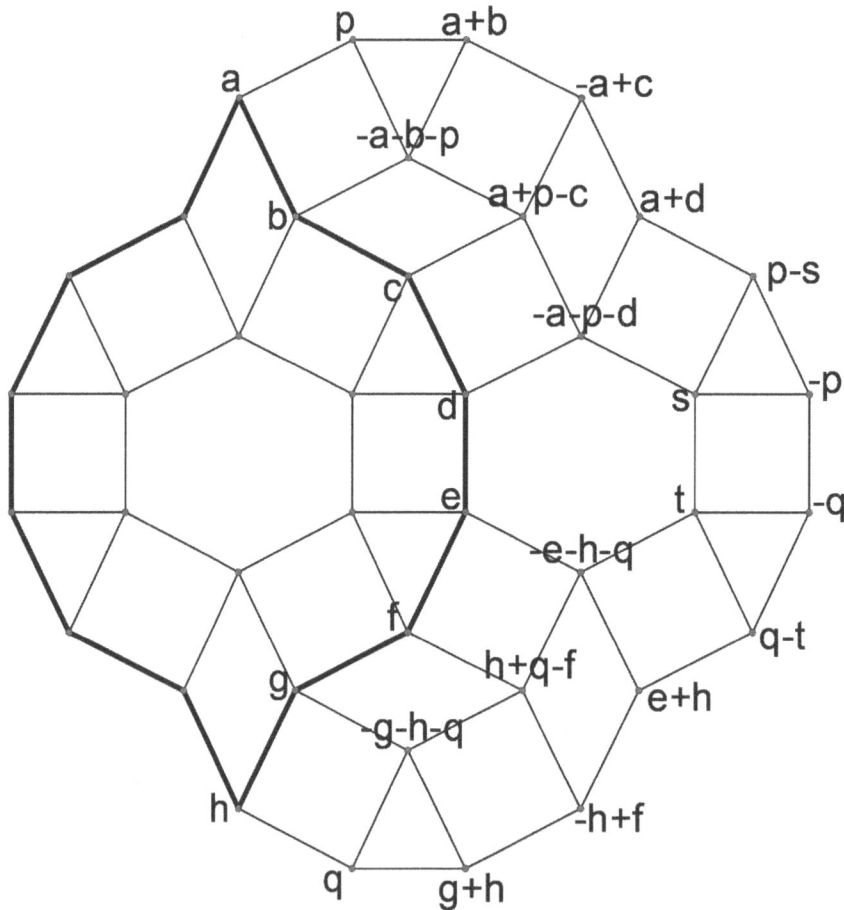

Figure 6.2

Similar to Π_5 the plane Π_6 is easily seen to be infinite dimensional.

6.2 Scoring configurations

As in $\prod_5$, the basic scoring configurations consist of the square, (M), the rhomb (R), the triangle, (Δ), and the hexagon, (H). Larger scoring configurations correspond to polygons which are nilpotent in the $\prod_6$ plane. Some of these are illustrated in Figure 6.3.

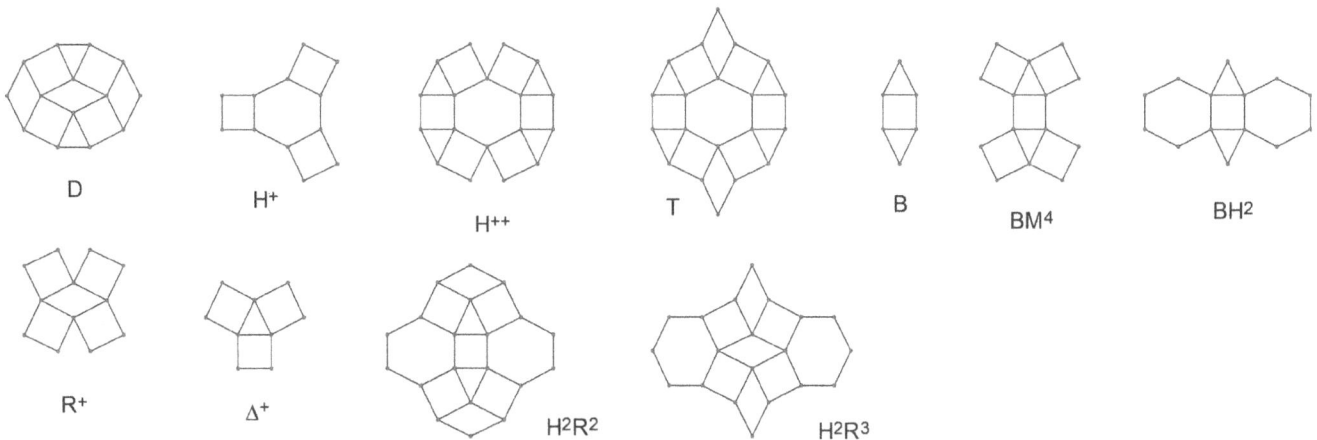

Figure 6.3

Any nilpotent structure with an axial or rotational symmetry is our suggestion for scoring configurations. Dead points occur much more frequently in the $\prod_6$ game than in $\prod_5$. We therefore recommend that blocking plays be disallowed. (Recall that a blocking play occurs when a nonscoring disc is played which creates an adjacent dead point.) Also playing a nonscoring disc on a boundary point belonging to a region which contains a dead point, should be disallowed.

6.3 The play

With the exception of the disallowance of blocking plays, the rules for play are the same as in the $\prod_5$ game. The polygon outlined in bold in Figure 6.1 is a barrier. As usual the boundary and interior of the barrier must be covered with discs and dead points before play outside the barrier is allowed.

CHAPTER 7

The $\prod_7$ game

7.1 The $\prod_7$ plane

In Figure 7.1 a section of the $\prod_7$ plane is illustrated.

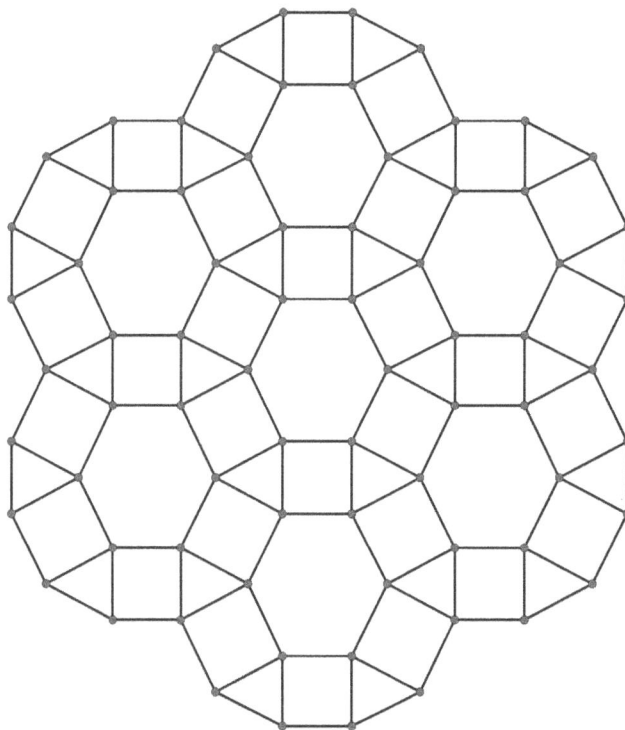

Figure 7.1

Our first and simplest game. As in the $\prod_5$ and $\prod_6$ games, the square, the rhomb, the triangle and the hexagon are the basic scoring configurations. With the exception of the prominently featured dodecagon (DD), the familiar scoring configurations B_2,

H^+, Δ^+ and $B_2{}^+$ occur also in the Π_7 plane. The Π_7 plane has a free locally nilpotent assignment and is infinite dimensional.

Unlike in Π_6 where only the boat B_2 occurs, *multiple boats* B_3, B_4, B_5, ... constitute a main scoring feature of this game. The notation B_m denotes a configuration consisting of connected complex of boats with m denoting the number of triangles in the configuration. The dodecagon is difficult to achieve so it should count more than just the corresponding boat configuration B_6. We suggest therefore that the value accruing from the dodecagon be counted twice; once as a dodecagon and again as a part of any connected boat configuration which includes the dodecagon.

7.2 The play

This game can be played in two versions: one with blocking and one in which blocking is not allowed. The scoring configurations described above work well with blocking allowed. If blocking is not allowed then larger scoring configurations are likely to occur. Two of these, H^3 and H^4 are illustrated in Figure 7.2.

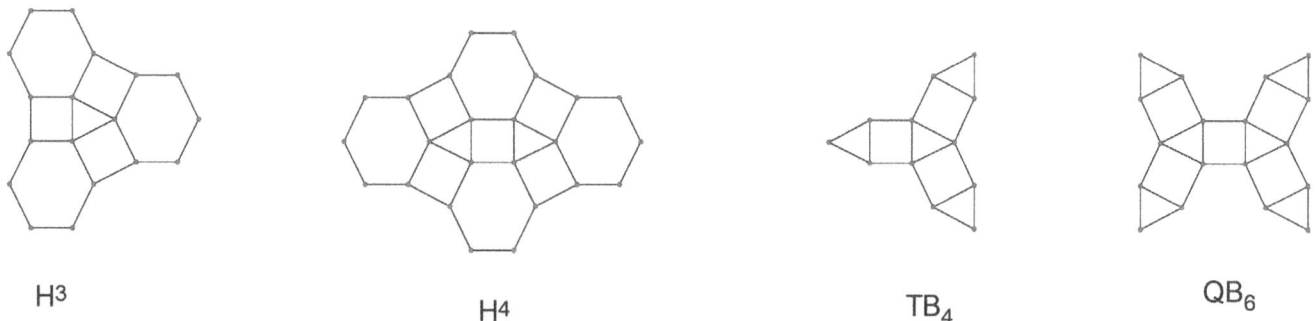

H3   H4   TB₄   QB₆

Figure 7.2

When blocking is not allowed, only the boat complexes TB_4, the *tripod boat complex,* and QB_6, the *quadruped boat complex* should be allowed as scoring positions. This is because large and unwieldy boat complexes are likely to occur. Even larger nilpotent structures can occur when blocking is not allowed, in particular structures involving multiple dodecagons. In our opinion these should not be allowed as scoring configurations; only DD, H^3 and H^4 should be allowed as large scoring structures.

Playing a nonscoring disc in a region occupied by a dead point should always be disallowed.

The play is the same as in the other games. The boneyard consists of 70 discs numbered {0, …, 9) in seven groups. Each player draws 10 discs from the boneyard and one more disc is drawn from the boneyard and placed on a point of the central hexagon, etc. No barrier is required in this game.

CHAPTER 8

The game $\prod_8$

8.1 The $\prod_8$ plane

A section of the $\prod_8$ tiling of the plane is illustrated in Figure 8.1.

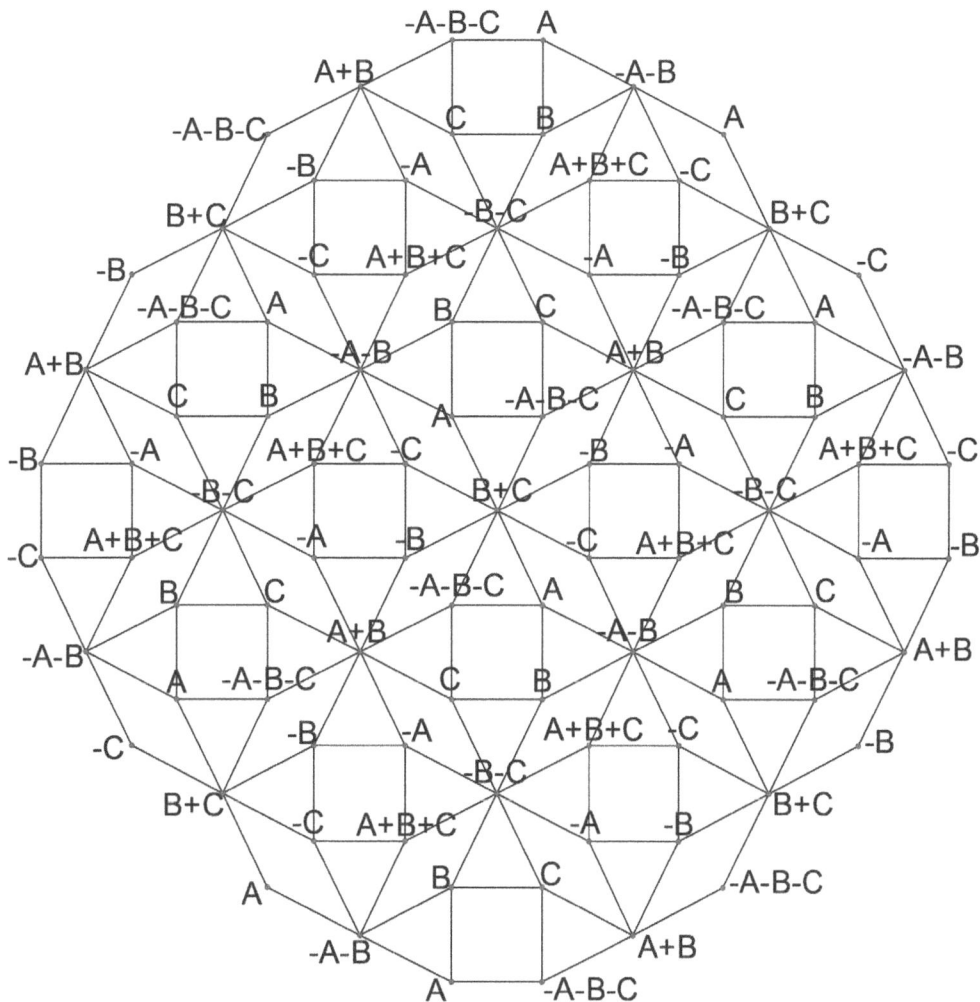

Figure 8.1

From Figure 8.1 it can be seen that the integers A, B, C generate the entire locally nilpotent assignment and that this is a free assignment. The dimension of Π_8 is thus 3. The basic tiles, (scoring configurations), consist of the square, (M), the equilateral triangle, (Δ), and the rhomb, (E), with interior angles 30 and 150 degrees.

8.2 Scoring configurations

Scoring configurations in the Π_8 game will, as usual, correspond to nilpotent Π_8 polygons. From Figure 8.1 various 0-sum point couples can be discerned. These are indicated in Figure 8.2.

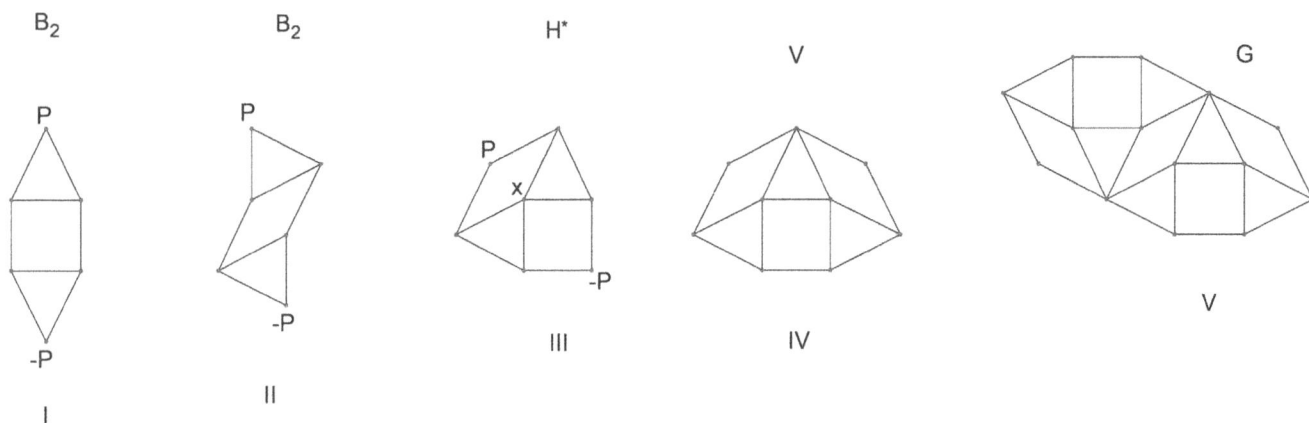

Figure 8.2

Figure 8.2(I,II) are composite with opposite points of the constituent triangles forming a 0-sum point couple. The 0-sum point couple {P, -P} occurs at opposite corners of the conditionally nilpotent hexagon H*. The center point x must be 0 for H* to have a 0 value. (We use an asterisk to indicate conditional nilpotence.) The 7-gon V and the semiregular Π_8 octagon G are intrinsically nilpotent; 0-sum point couples can be used to demonstrate this.

Some more scoring configurations are illustrated in Figure 8.3. Configurations M+, E+, O are prime and the remaining structures in Figure 8.3 are conditionally nilpotent: A+B must equal 0 for O* to have 0 value; for P*, S* and DD* to have 0 value, the center point x must be 0.

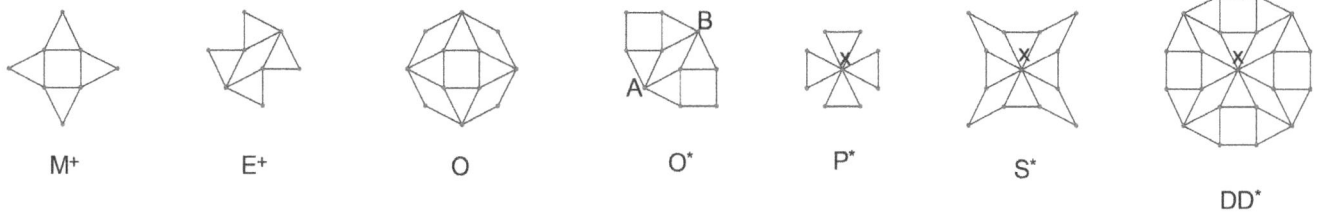

Figure 8.3

The scoring configurations in Figures 8.2, 8.3 are the most likely to occur in the play of the $\prod_8$ game. It is nevertheless interesting to consider what other convex $\prod_8$ polygons might be nilpotent or conditionally nilpotent. The large dodecagon, (DD) in Figure 8.4 is nilpotent. The octagon (O**), and the dodecagon (DD**) in Figure 8.4 are conditionally nilpotent: both require A+B = 0 to have 0-value. We leave the proof of these facts as an exercise.

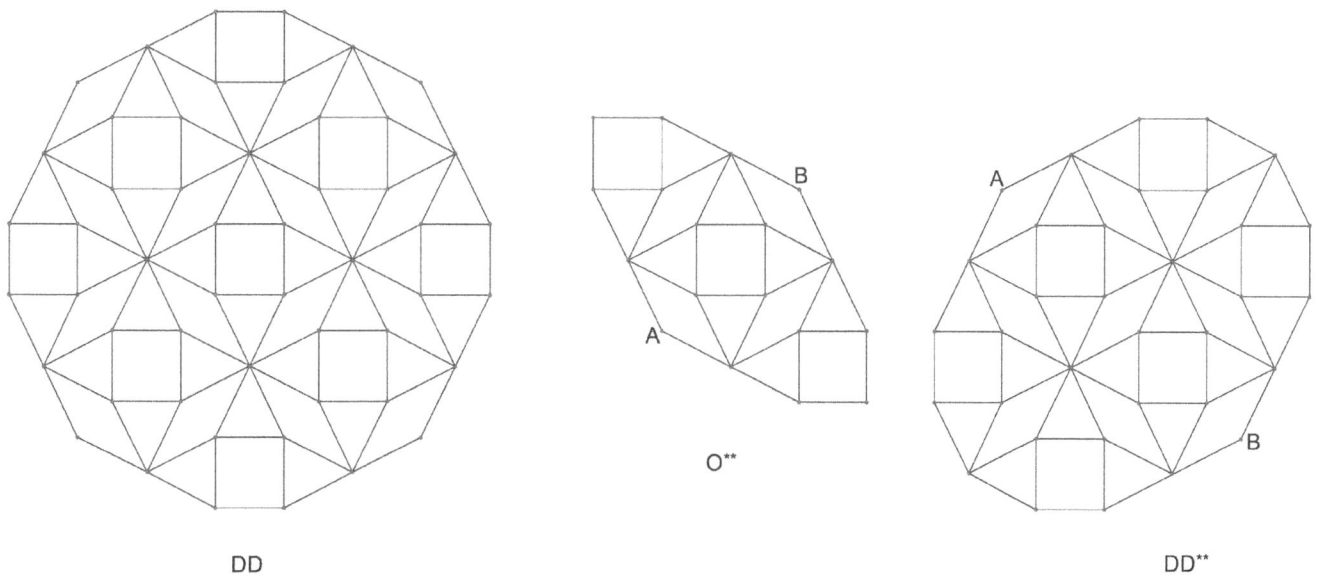

Figure 8.4

Two more intrinsically nilpotent $\prod_8$ polygons are illustrated in Figure 8.5. In Figure 8.5(I) the bold polygon is nilpotent as well as the extension indicated by the dashed lines. Figure 8.5(II) is a dashed extension of the octagon O.

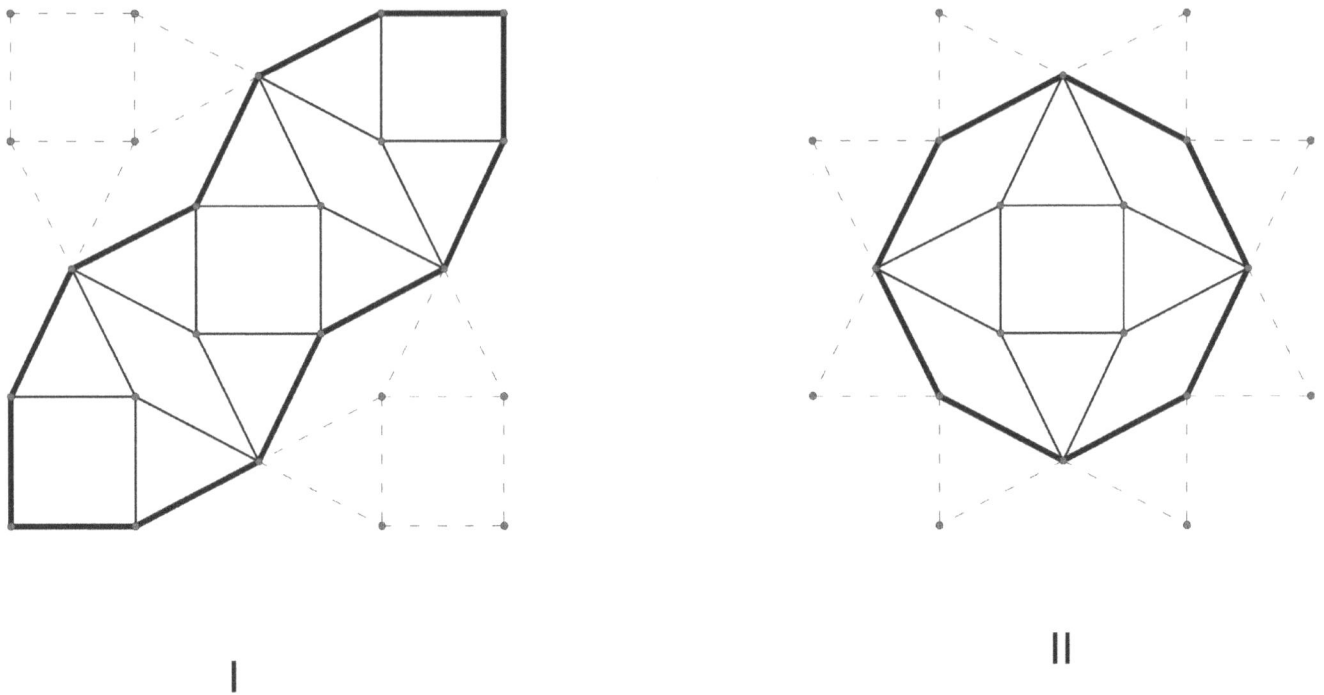

I

II

Figure 8.5

8.3 Rules for the $\prod_8$ game

The board for the $\prod_8$ game is illustrated in Figure 8.6. It has 136 points and the game is provided with 140 numbered discs in 14 groups of {0, …,9}. A barrier is indicated in bold in Figure 8.6. Each player draws 10 discs from the boneyard, and an additional disc is drawn and placed on one of the points of the central square.

As usual, in the game nilpotence over Z_5 is substituted for nilpotence over Z. Thus scoring configurations must have a value which is a multiple of 5.

The $\prod_8$ game is best played with blocking disallowed, as dead points occur in sufficient quantity from the placement of scoring discs. Also placing a nonscoring disc in a region occupied by a dead point should be disallowed. Otherwise the rules of play are the same as in our previous games.

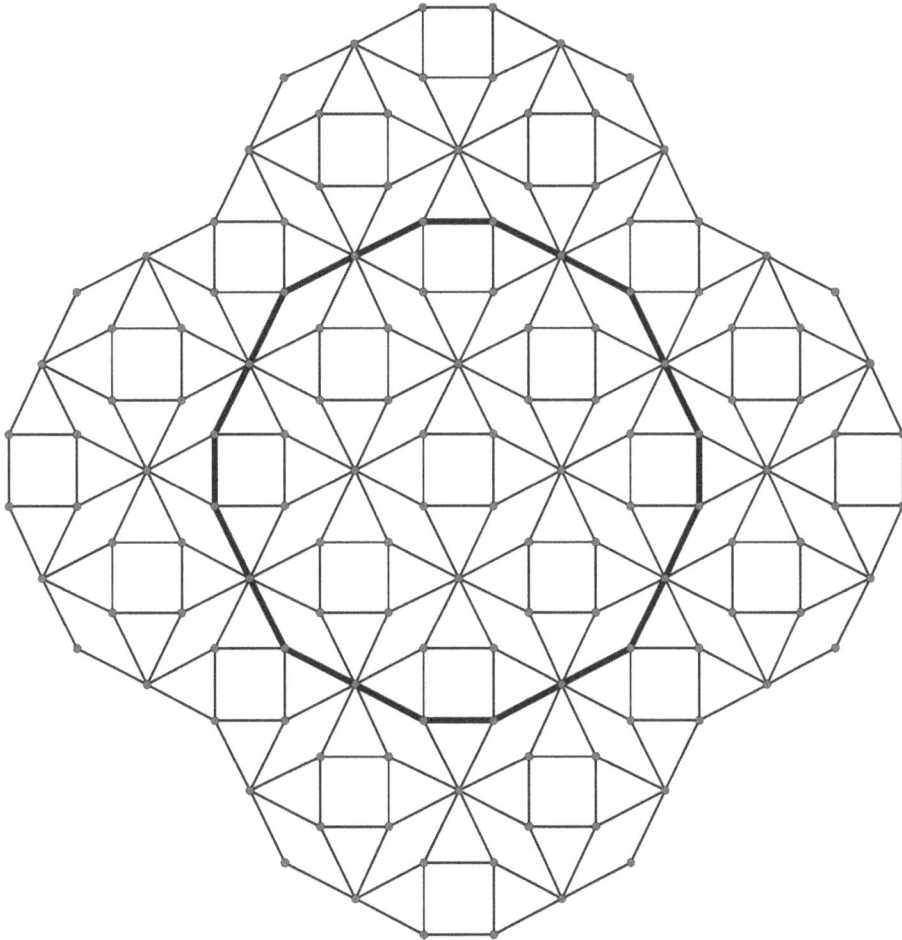

Figure 8.6

CHAPTER 9

Nilpotent polycubes

9.1 Space tiling with cubes

In this Chapter a 3 dimensional game will be created based on a covering of 3 space with cubes. If constructed in 3 dimensions the basic scoring configuration will consist only of the unit square (M(2,2)). When represented conveniently in the plane, the basic scoring configurations take the shape of rectangles, rhombs and parallelograms. In Figure 9.1 the game board is illustrated. It is a section of a *space tiling by* cubes which consists of a 4 by 4 by 5 box. (In this Chapter we measure the length of a side by the number of vertices it contains.)

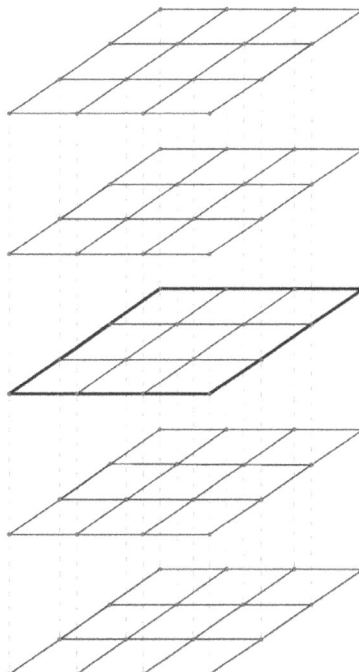

Figure 9.1

The game board consists of 5 parallel planes each a square with 16 vertices. Play starts in the central plane which constitutes a barrier.

Let us denote by Γ the tiling of 3-space by cubes. Γ is infinite dimensional in our sense; i.e., it takes an infinite generating set from Z to create a locally nilpotent assignment. This assignment is free as a consequence of the following observation.

Theorem 9.1 *If seven points of a cube are assigned integer values so that all but three faces have 0 sum, then there exists an integer value which can be assigned to the eighth point which completes a 0 -sum for the remaining three faces.*

9.2 Nilpotent rectangles

There are two types of nilpotent structures in Γ: (i) those which lie in the same plane and (ii) those which do not lie in one plane. For those whose vertices lie in a single plane we will just consider rectangular polysquares $M(a,b)$ where a, b denote the number of points in two adjacent sides. In case a, b are both even then $M(a,b)$ has an $\{M(2,2)\}$ – dissection, as in Figure 9.2(I), and thus is nilpotent. If one of the parameters, say a, is odd, then $M(a,b)$ is nonnilpotent. This is easily seen, as in the example in Figure 9.2(II). A nonzero valued locally nilpotent assignment is created by just alternating 1's and -1's along an odd length segment.

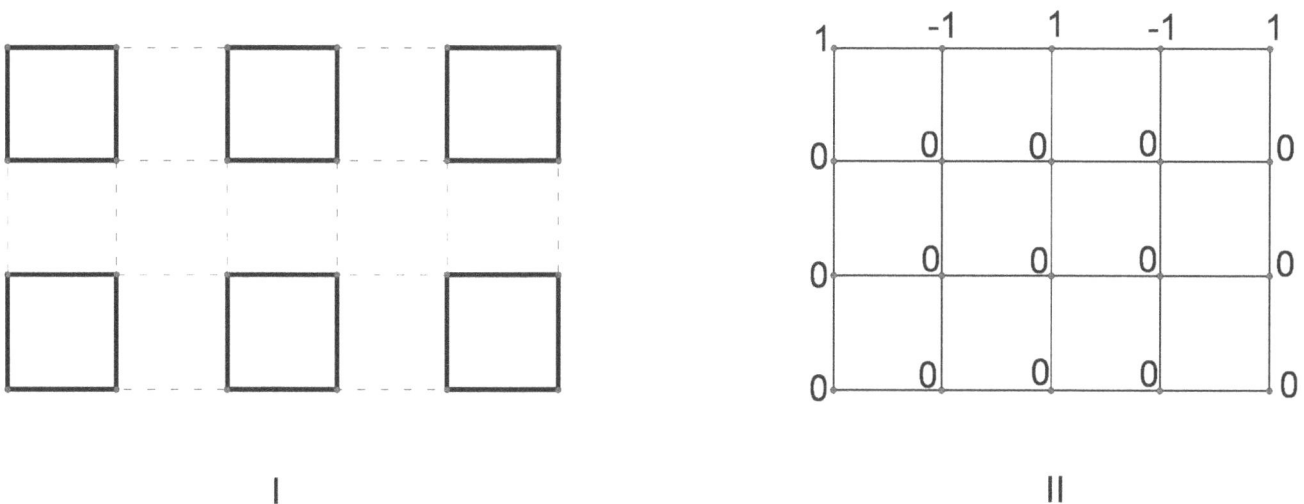

Figure 9.2

9.3 Nonplanar nilpotent structures in Γ

Among all the possible nonplanar nilpotent Γ configurations, we will consider only *boxes*. A box B(a,b,c) is a lattice in Γ consisting of a stack of c planes each of which is an M(a,b) rectangle. In case 2 of the parameters are even as in Figure 9.3(I), then B(a,b,c) has an {M(2,2)} – dissection and thus is nilpotent.

In case 2, (or 3), of the parameters are odd then B(a,b,c) is nonnilpotent. If, for example, a, b are odd then any of the rectangles M(a,b) which constitute the box can be given an alternating 1, -1 assignment, and all other points can be assigned the value 0. A locally nilpotent assignment results which has a nonzero value, and thus B(a,b,c) is nonnilpotent. The box B(3,3,4) is illustrated as an example in Figure 9.3(II).

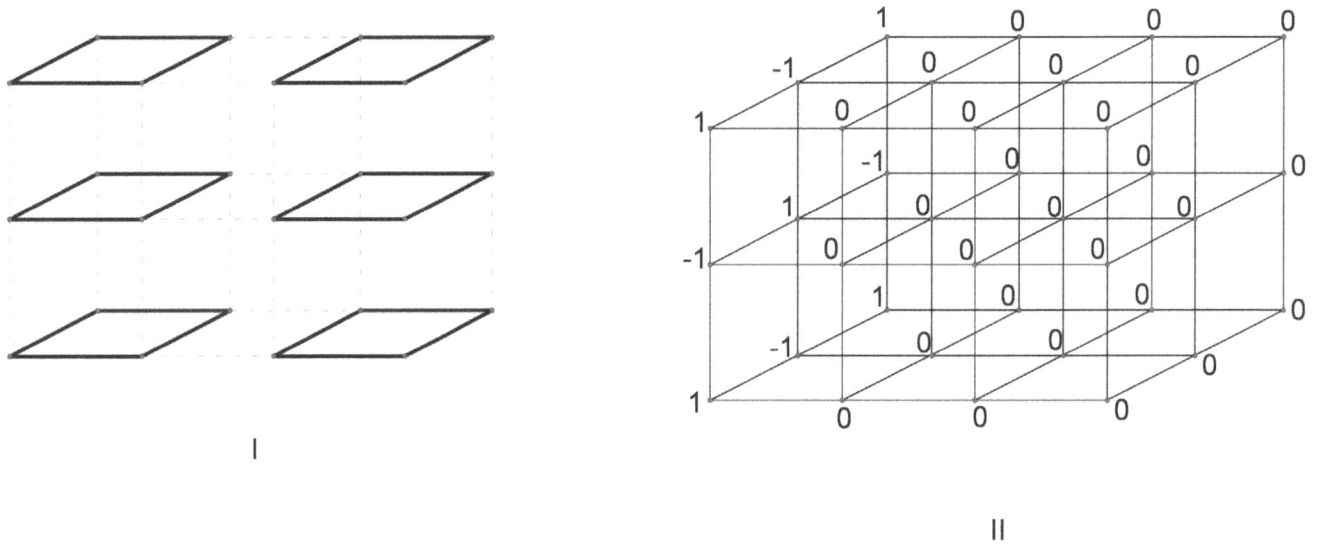

Figure 9.3

9.4 Conditional nilpotence for Γ rectangles and boxes

In Figure 9.4 the rectangle M(3,3) is illustrated.

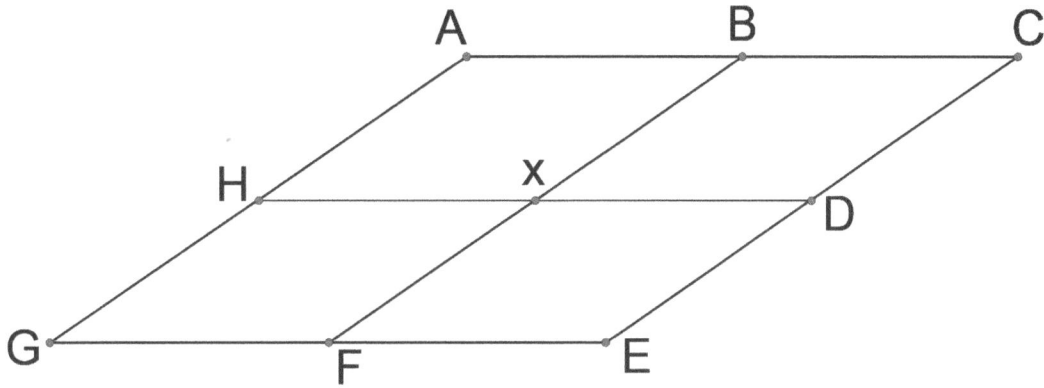

Figure 9.4

Conditional nilpotence for M(3,3) can be achieved as follows.

Theorem 9.2 *With regard for the labeling in Figure 9.4, M(3,3) has 0 value in any locally nilpotent assignment iff A+E = x. In this case G+C = x also holds.*

As usual we leave the proof of the Theorems as exercises. Note that, less elegantly, the value of M(3,3) is 0 iff C+D+E+F+G = 0. If a, b are odd then this less elegant approach can be always be used; i.e., M(a,b) = 0 iff the sum of the rightmost column and the remaining values of the bottommost row is 0. This is because the section M(a-1, b-1) has a cover with M(2,2) squares.

In case one of the parameters, say a, is even and b is odd, then the section M(a, b-1) has a cover with M(2,2) squares. This leaves only the rightmost column uncovered. Thus M(a,b) = 0 iff the sum of the values in this column is 0.

For the box B(3,3,3) we have a result similar to Theorem 9.2.

Theorem 9.3 *With reference to Figure 9.5, B(3,3,3) has 0 value in any locally nilpotent assignment in case A+G = x. In this case K+E = x also holds.*

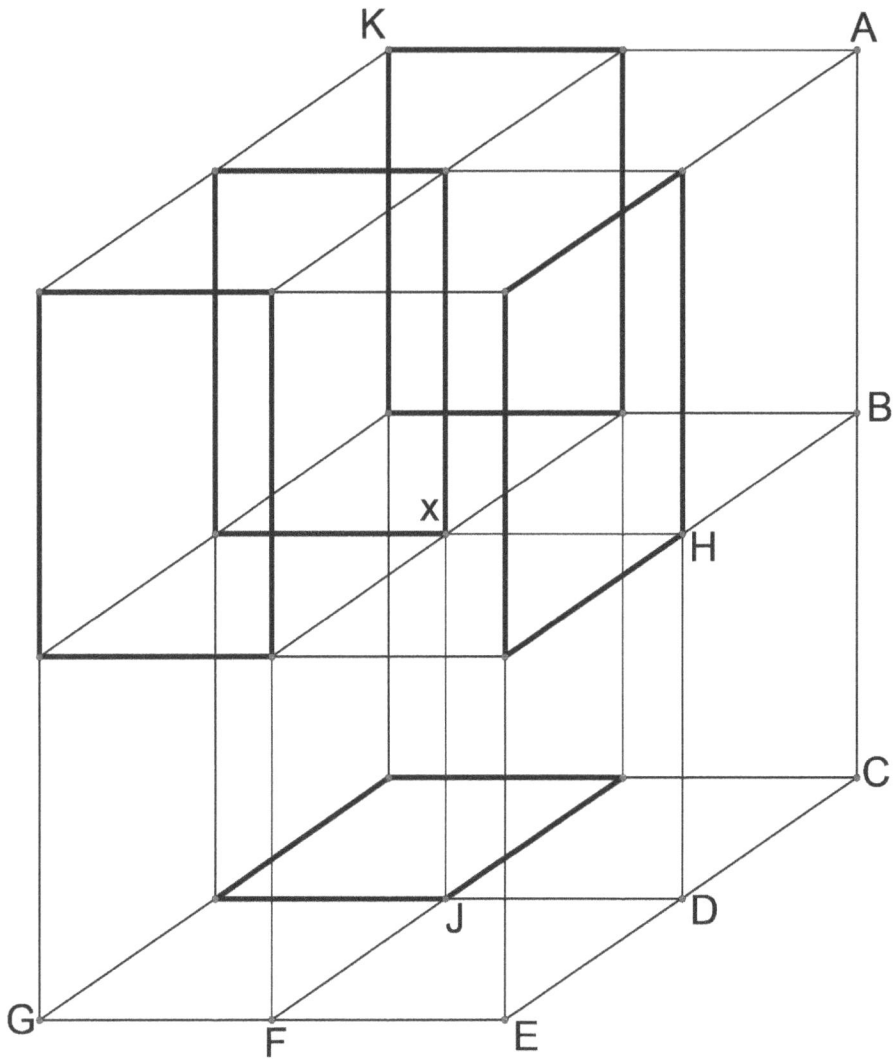

Figure 9.5

The partial cover by M(2,2) squares shows that B(3,3,3) = 0 also iff A+B+C+D+E+F+G = 0. In case a,b,c are all odd, then this method of summing the values on the right side and bottom boundaries can always be used to determine conditional nilpotence.

In case two of the parameters, say a, b are odd and c is even as in the example in Figure 9.6, then all except one even length row has an M(2,2) cover. Thus B(a,b,c) = 0 iff the sum of this uncovered row is 0.

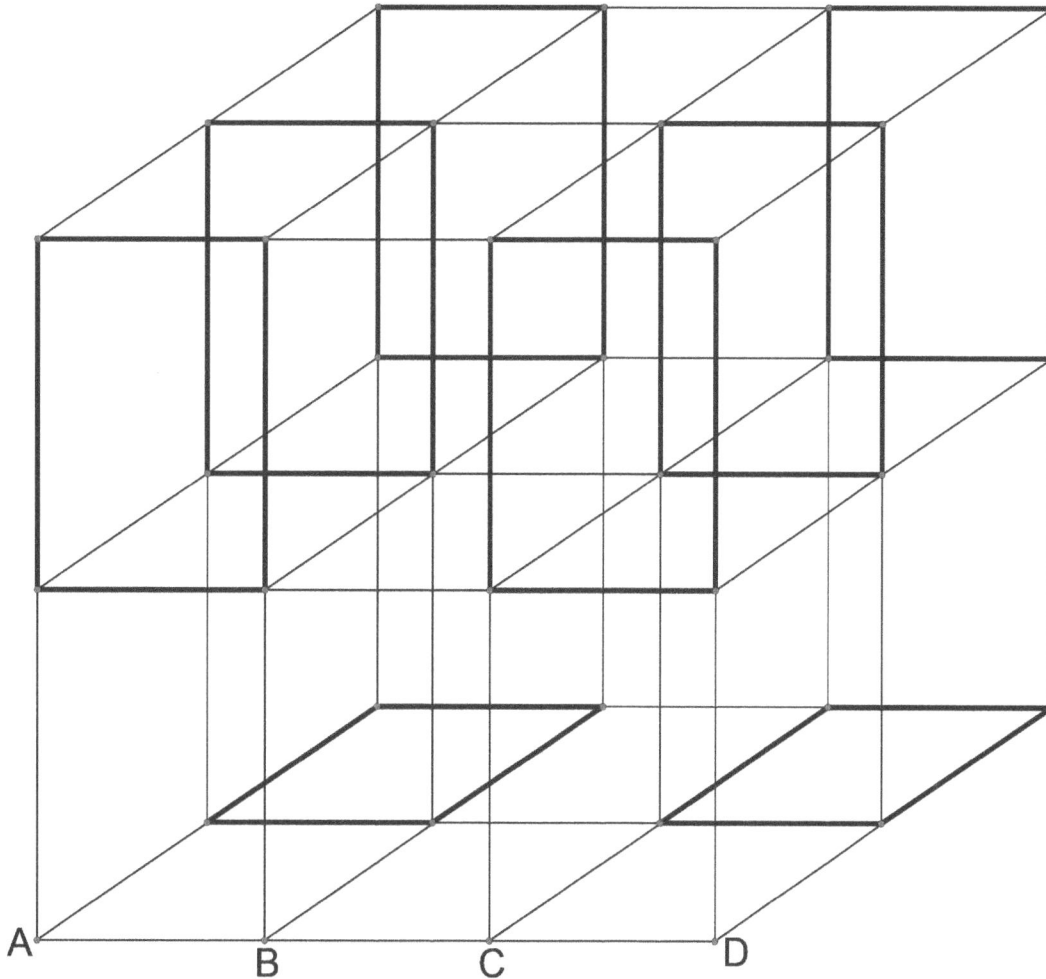

Figure 9.6

9.5 The Γ game

As already stated, the playing board is as in Figure 9.1. The central plane is a barrier; it must be covered either by numbered discs or dead points before points on either of the two adjacent planes can be covered.

M(2,2) is the basic scoring configuration. There are two ways to further define scoring configurations:

(1) Two squares in the same plane which cover an M(2,4) rectangle; four mutually nonadjacent squares which lie in the same plane; a B(2,2,2) cube; a series of adjacent squares which cover a box B(2,2,c), (c can equal 2,3,4 or 5); any box which has an M(2,2) cover.

(2) The single play of a numbered disc can complete 1, 2, or 3 squares. The score for the play is just the sum of the scores determined by each of the created squares. This scoring method has the advantage of simplifying tabulation.

The rules of play are essentially the same as our previous games: each player selects 10 discs from the boneyard and one more disc is selected and played in the central plane. In this game there are a total of 80 points, and thus there are 80 discs in 8 groups of (0, ..., 9). Rules for scoring and nonscoring plays as well as tabulation of double scores are as usual. Dead points occur frequently during the natural course of play, therefore we recommend that blocking plays be disallowed, and also playing a nonscoring disc in a region occupied by a dead point should be disallowed.

CHAPTER 10

Nilpotent Polygons in the Π_9 Plane

10.1 The Π_9 plane

The playing board for the Π_9 game is illustrated in Figure 10.1. The Π_9 plane corresponds to one of the 24 types of *proper* tilings by pentagons, (see [3]). A tiling is proper in case adjacent tiles intersect in a common edge.

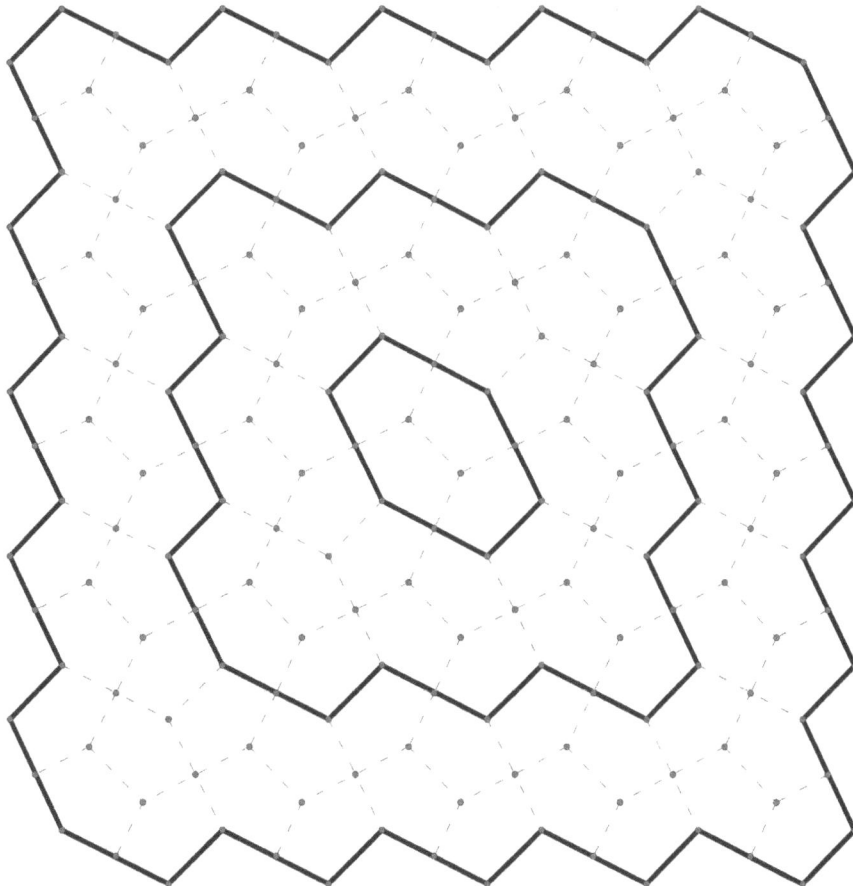

Figure 10.1

Unlike our previous planes, dead points are inevitable, even over the most careful attempt at a locally nilpotent assignment over Z, (or even over Z_5). Thus a free locally nilpotent assignment does not occur for the $\prod_9$ plane. The basic scoring configurations consist of the pentagon, (P), and the tripod, (T). The plane consists of H = H(1,2,2,1,2,2) hexagons which occur in two orientations.

10.2 Inlets and bays

An *inlet* is illustrated in Figure 10.2(I), and a *bay* is illustrated in Figure 10.2(II). These are special $\prod_9$ formations which occur frequently on the perimeter of the growing central configuration during the play of the game.

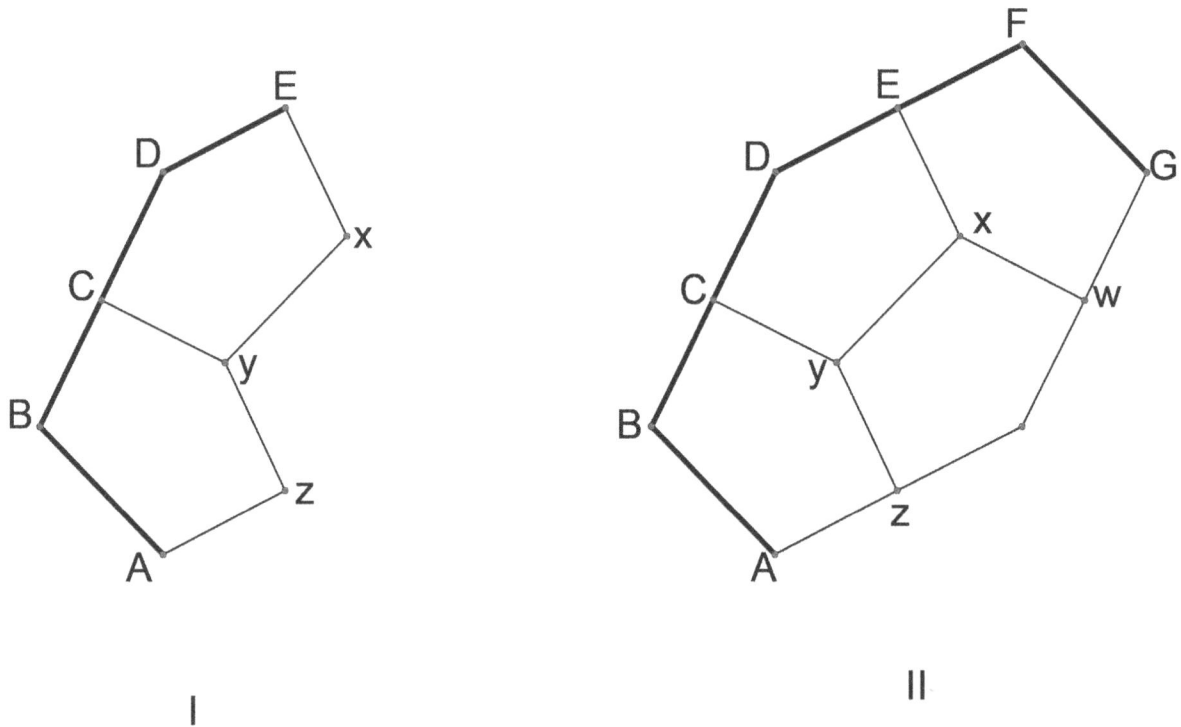

Figure 10.2

For the inlet consisting of the played points A,B,C,D,E in Figure 10.2(I) the equations

(1) A+B+C+y+z = 0 mod 5
(2) C+D+E+x+y = 0 mod 5
(3) C+x+y+z = 0 mod 5

must be satisfied. The solutions are A+B = x mod 5; E+D = z mod 5, and y = -A-B-C-D-E. Thus x, y, z have unique solutions mod 5. The easiest one to see is x = A+B since x is opposite to side AB of a pentagon. *In the game we require that the point x be the first of the three points x,y,z to be played in the inlet and that the played value must equal A+B mod 5.*

It is legal but it doesn't make much sense to play at point x without at least having the required value for y in hand. *For this reason, and only in this case, a player is allowed to play the nonscoring x and the value for y in the same turn without the turn being completed.* This is called a *double play.*

The bay consisting of the played points A,B,C,D,E,F,G in Figure 10.2(II) is a formation which contains two inlets. A player may choose between playing x = A+B and attempting to complete one of the inlets, and y = F+G, in an attempt to complete the other inlet. If the inlet with points x,y,z is successfully completed, then in most cases the point w becomes dead.

10.3 Scoring configurations

Besides the basic scoring configurations, (T,P), the hexagon H often occurs with 0 value (mod 5). This happens when the opposite boundary points of the two central pentagons have 0 sum. Whether or not H has a 0 value depends upon the order in which its values are assigned. Three conditionally nilpotent Π_9 polygons are illustrated in Figure 10.3. The requirement is that the central point have value 0 or 5.

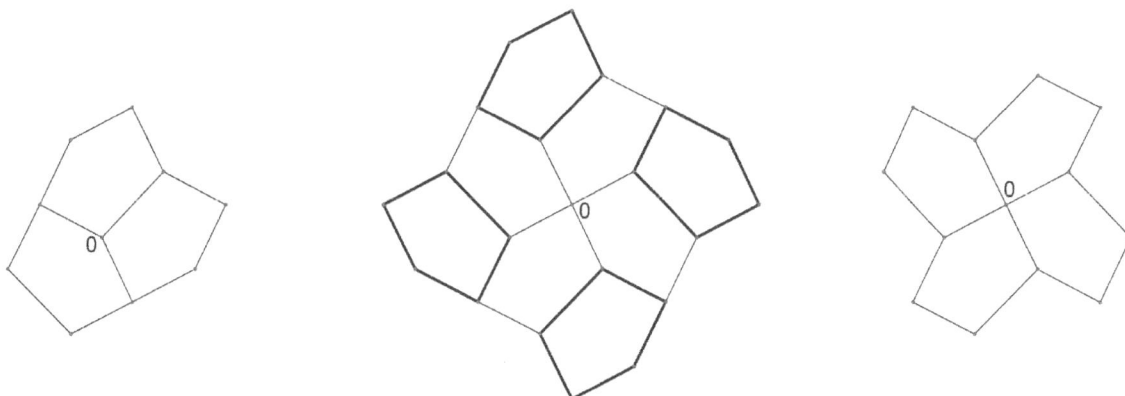

Figure 10.3

Larger nilpotent polygons can be created by amalgamating 0 value hexagons, but our preference is to restrict scoring configurations to T, P, H and the three conditionally nilpotent polygons in Figure 10.3.

10.4 The play

The playing board, (Figure 10.1), has 180 points. Usually there are a significant number of dead points and unplayed or unplayable points, so a total of 160 numbered discs should be adequate. These occur in 16 groups of $\{0, \ldots, 9\}$.

As in all of our games, each player draws 10 discs from the boneyard and then one additional disc is drawn and placed on one of the two central points in the center hexagon. The central hexagon constitutes the first barrier, and the second barrier, indicated in bold in Figure 10.1, contains 9 hexagons in one orientation and 4 in the other orientation.

Blocking plays and playing a nonscoring disc in the same basic region as a dead point are not allowed. The rules for double scores and the play of a sequence of scoring discs followed by a nonscoring disc are the same is in the previous games.

CHAPTER 11

Derivative planes

11.1 The derivative graph

Let $\prod$ represent a tiling of the plane with convex polygons. Now construct a new tiling of the plane by putting a point on the midpoint of every edge of $\prod$, and then joining two of these points if they correspond to adjacent edges of a tile of $\prod$. The result is again a tiling of the plane with convex polygons, which we call the *derivative* of $\prod$, and denote by $D(\prod)$. The construction can be repeated to obtain the *second derivative* $D^2(\prod)$ of $\prod$. In Figure 11.1 the derivatives of the hexagon lattice $\prod_2$ are illustrated.

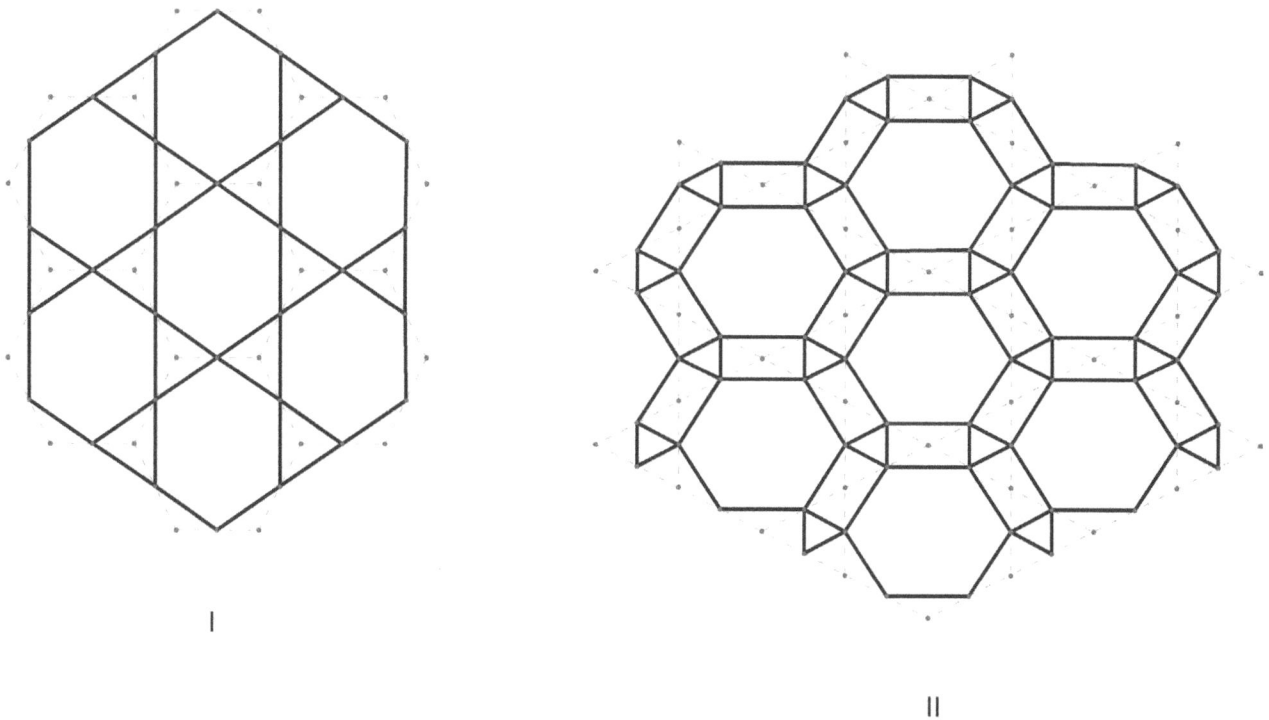

I   II

Figure 11.1

From Figure 11.1 we see that $D(\prod_2) = \prod_1$, and from Figure 11.1(II) we see that $D^2(\prod_2) = \prod_7$. In Figure 11.2 the further derivative $D(\prod_7) = D^3(\prod_2)$ is illustrated. Note that $D(\prod_7)$ is *isomorphic to* $\prod_5$. Two graphs are isomorphic if they are essentially the same; they differ only in the way they are drawn,

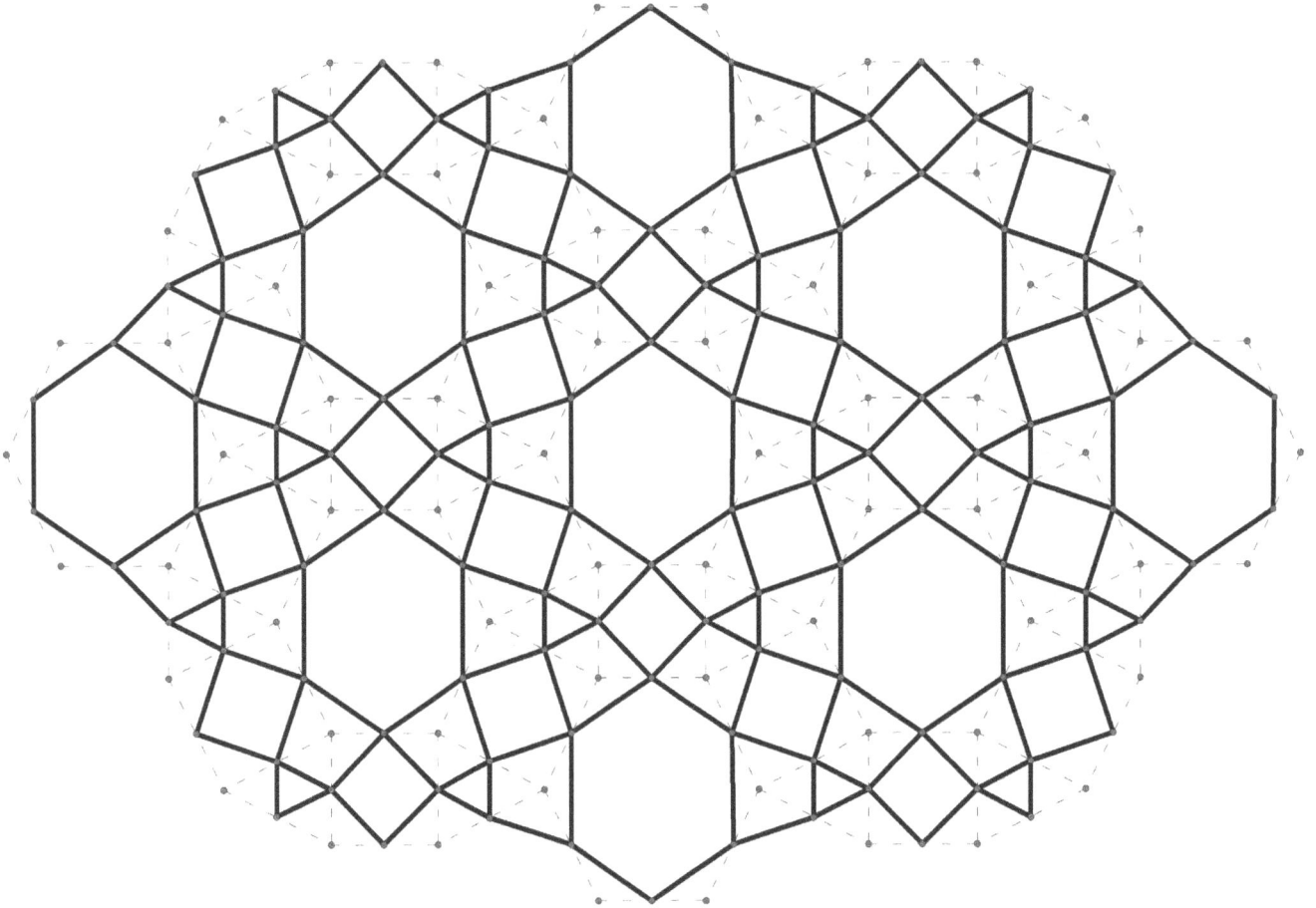

Figure 11.2

Suppose P is a nilpotent polygon in a plane $\prod$ all of whose basic regions are convex polygons. Construct a polygon $D(P)$ in $D(\prod)$ by (i) drawing the derivative $D(R)$ for each basic region R in P, and (ii) including all edges in $D(\prod)$ which form a region containing a vertex of P. In this case we make the following Conjecture.

Conjecture 11.1 *Let $\prod$ denote a tiling of the plane whose basic tiles are convex polygons. If P is a nilpotent polygon in $\prod$, then D(P) is nilpotent in D($\prod$).*

In Figure 11.3 some evidence for Conjecture 11.1 is given.

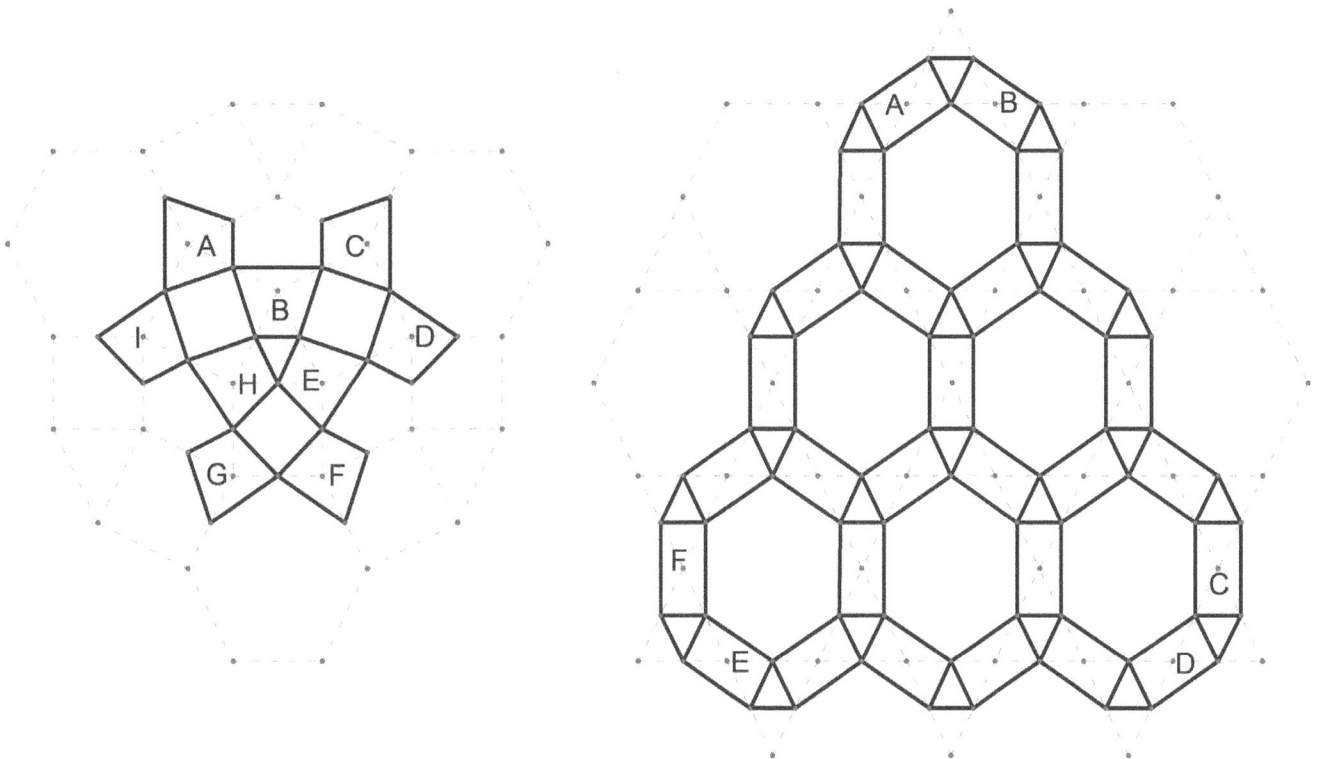

Figure 11.3

In the construction at left in Figure 11.3 the dashed base polygon P might be denoted by Δ_1^+ indicating the extension of Δ_1 by base polygons, (squares), attached to the sides. Its vertices have been labeled A, …, I. We call this a *side extension*. The derivative D(P), indicated in bold, is nilpotent as the reader may wish to show. If θ denotes any base polygon in a tiling $\prod$ of the plane in which every tile is a convex polygon, and θ^+ denotes a side extension of θ, then we claim that $D(\theta^+)$ is nilpotent in $D(\prod)$.

At right in Figure 11.3 the dashed H = H(1,5,1,5,1,5) hexagon with corners A, …, F is nilpotent in $\prod_1$. Its derivative in $D(\prod_1) = \prod_7$, indicated in bold is nilpotent in $\prod_7$.

11.2 The derivative of $\prod_9$

In Figure 11.4 a section of $\prod_9$ is dashed along with its derivative in bold.

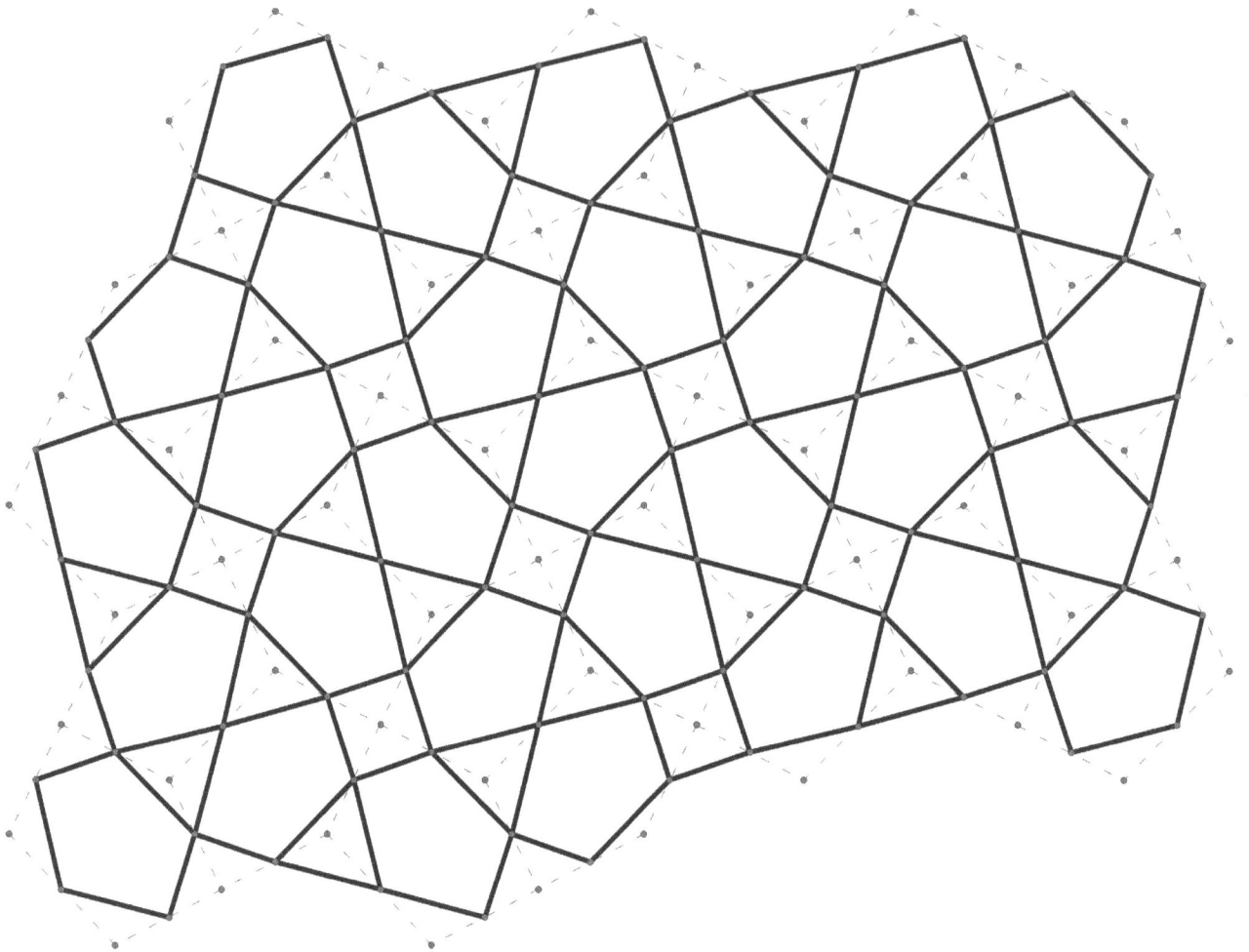

Figure 11.4

Let us denote $D(\Pi_9)$ by Π_{10}. In Figure 11.6 two of the key nilpotent polygons in Π_{10} are illustrated in bold. We call the nilpotent polygon in Figure 11.6(I) the *large square*, and the nilpotent polygon in Figure 11.6(II), the *happyman*. Note that neither the happyman nor the large square are derivatives of nilpotent polygons in Π_9. Another nilpotent Π_{10} polygon, the *propeller* is illustrated in Figure 11.5. It is also the derivative of a nonnilpotent polygon in Π_9.

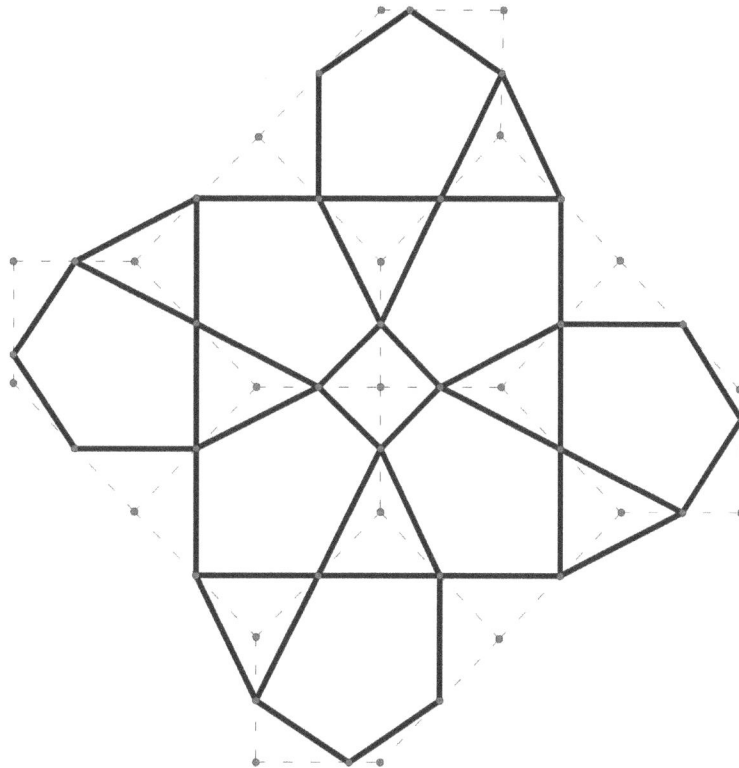

Figure 11.5

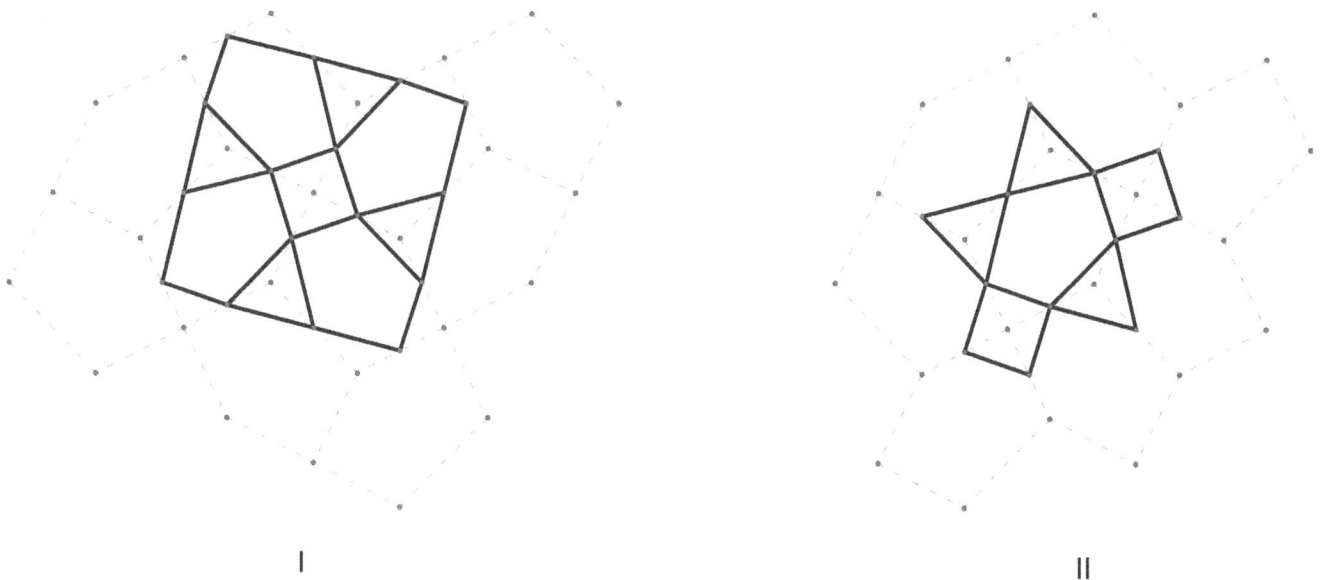

I

II

Figure 11.6

There are also a few conditionally nilpotent $\prod_{10}$ polygons which should be included as scoring configurations. These are illustrated in Figure 11.7.

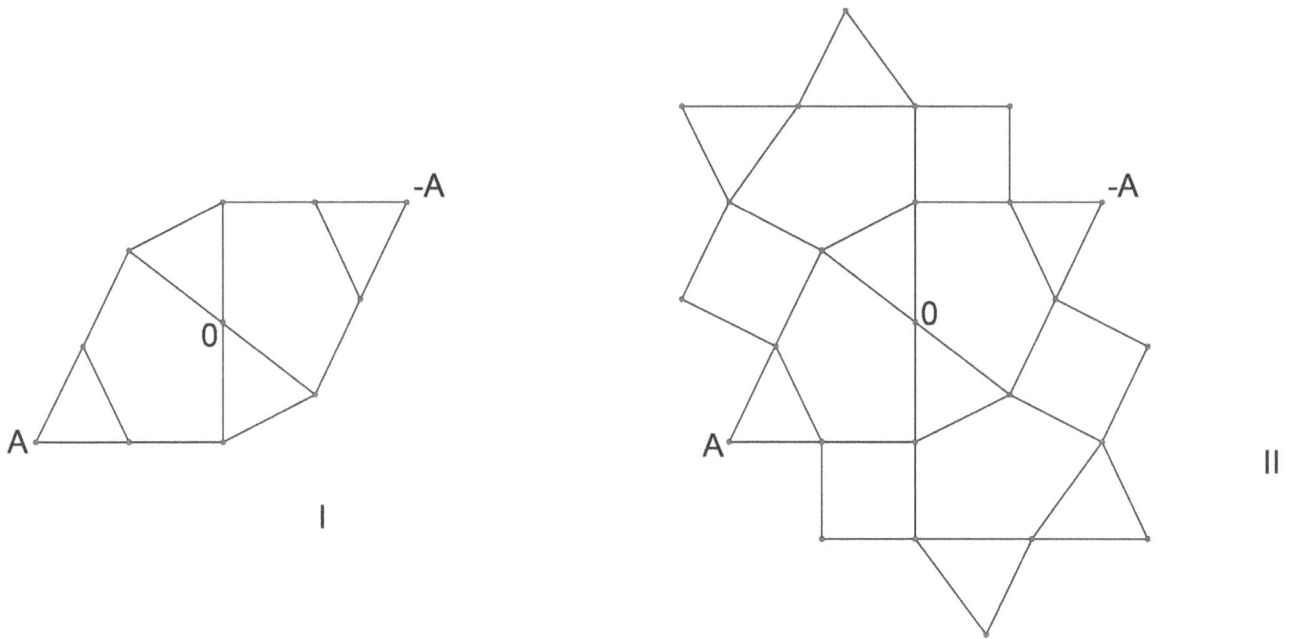

Figure 11.7

In both Figure 11.7(I) and 11.7(II) the requirement for nilpotence is that the central point have a 0 value as indicated. The game is played over Z_5 as usual, so the central points can also have a value of 5. It is an easy exercise to show that the point couples {A, -A} must have a 0 sum in any locally nilpotent assignment. Thus the octagon obtained by deleting the points A, -A in Figure 11.7(I) is also conditionally nilpotent. Similarly, the polygon obtained by deleting the point couple {A,-A} in Figure 11.7(II) is also conditionally nilpotent.

11.3 The $\prod_{10}$ game

Composite nilpotent $\prod_{10}$ polygons are plentiful, but in our opinion they should not count as scoring configurations. The basic triangle, small square and pentagon along with the large square, happyman, the propeller and the conditionally nilpotent configurations in Figure 11.7 should constitute the full set of scoring configurations. The existence of prime $\prod_{10}$ polygons larger than the propeller we leave as an open question.

The playing board for the $\prod_{10}$ game is illustrated in Figure 11.8. In this rendition not all the large squares are square and the happymen have one arm smaller than the other. This, of course, makes no difference to the game, since it is only important that the various graphs for Π_{10} are *isomorphic*. Most of our renditions of the base planes for our games are constructed with vertices at the points of a square lattice. This eases construction using Geometers Sketch Pad since this software provides a square lattice.

The central large square is a barrier and blocking should not be allowed. Also no nonscoring disc is allowed to be placed in a basic region containing a dead point. Otherwise the rules for play are the same as our previous games. There are 112 points on the board and the boneyard consists of 110 discs in 11 groups of {0, …, 9). Players draw 10 discs from the boneyard to constitute their hands. One more disc is drawn from the boneyard and placed on one of the four points of the central small square, etc.

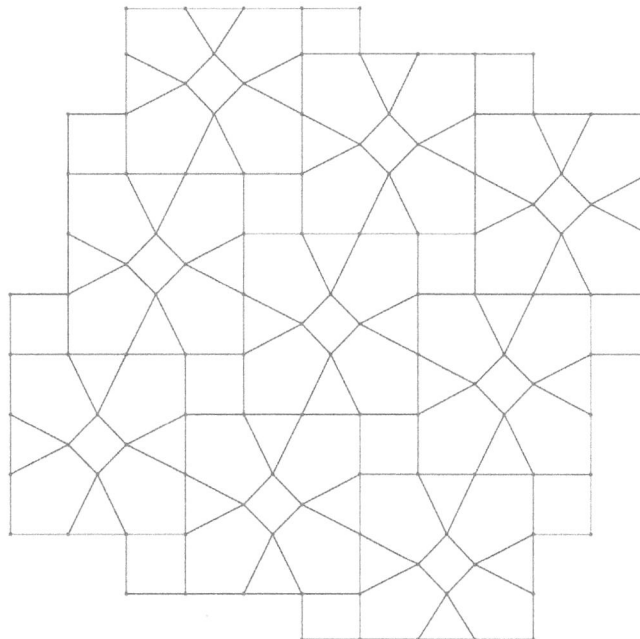

Figure 11.8

10.4 The derivative of Π_8

A graph isomorphic to a section of the derivative of Π_8 is illustrated in Figure 11.9. We denote this graph by $D(\Pi_8) = \Pi_{11}$.

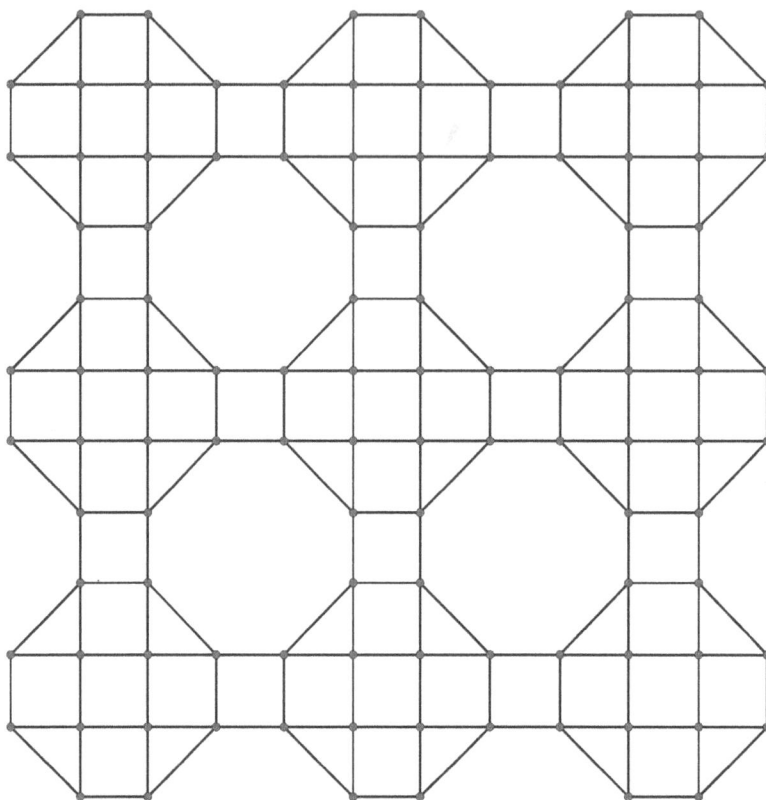

Figure 11.9

The Π_{11} plane is infinite dimensional and has a free locally nilpotent assignment over Z. Scoring configurations are numerous but for the Π_{11} game we suggest that these be confined to the ones illustrated in Figure 11.10. The right triangle, small square and *empty octagon* in Figure 11.10(I,II,III) are the basic scoring polygons. The *full octagon* in Figure 11.10(IV) and the remaining configurations in Figure 11.10 are composite. The large square in Figure 11.10(VIII) occurs rarely, and configurations VI, VII are optional. Figure 11.10(IX) is a linear polysquare consisting of 11 squares which occurs in the playing board, (Figure 11.9), in both horizontal and vertical orientations.

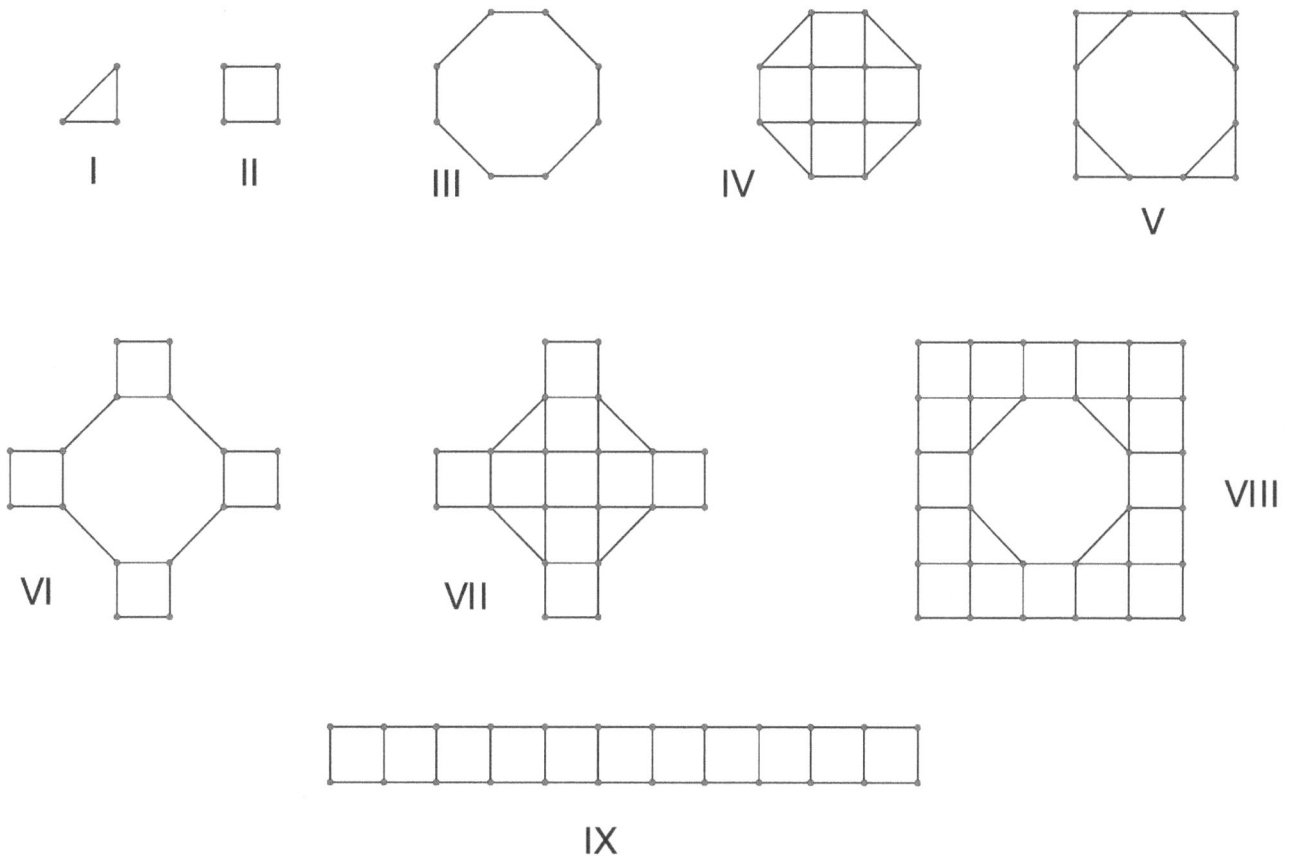

Figure 11.10

As far as the play is concerned, blocking should not be allowed. Also, playing a nonscoring disc in a tile containing a dead point should not be allowed. The playing board, (Figure 11.9), has 108 points. The number of playing discs is 110 in 11 groups of {0, ..., 9}. Each player draws 10 discs from the boneyard and then one more disc is drawn and placed on one of the points in the central small square. There are no barriers and play proceeds as in previous games.

CHAPTER 12

The Π_{12} game

12.1 The Π_{12} plane

In Figure 12.1 a section of the Π_{12} plane is illustrated. This section will be used as the playing board for the Π_{12} game. The basic scoring configurations consist of the large triangle, the small triangle and the pentagon.

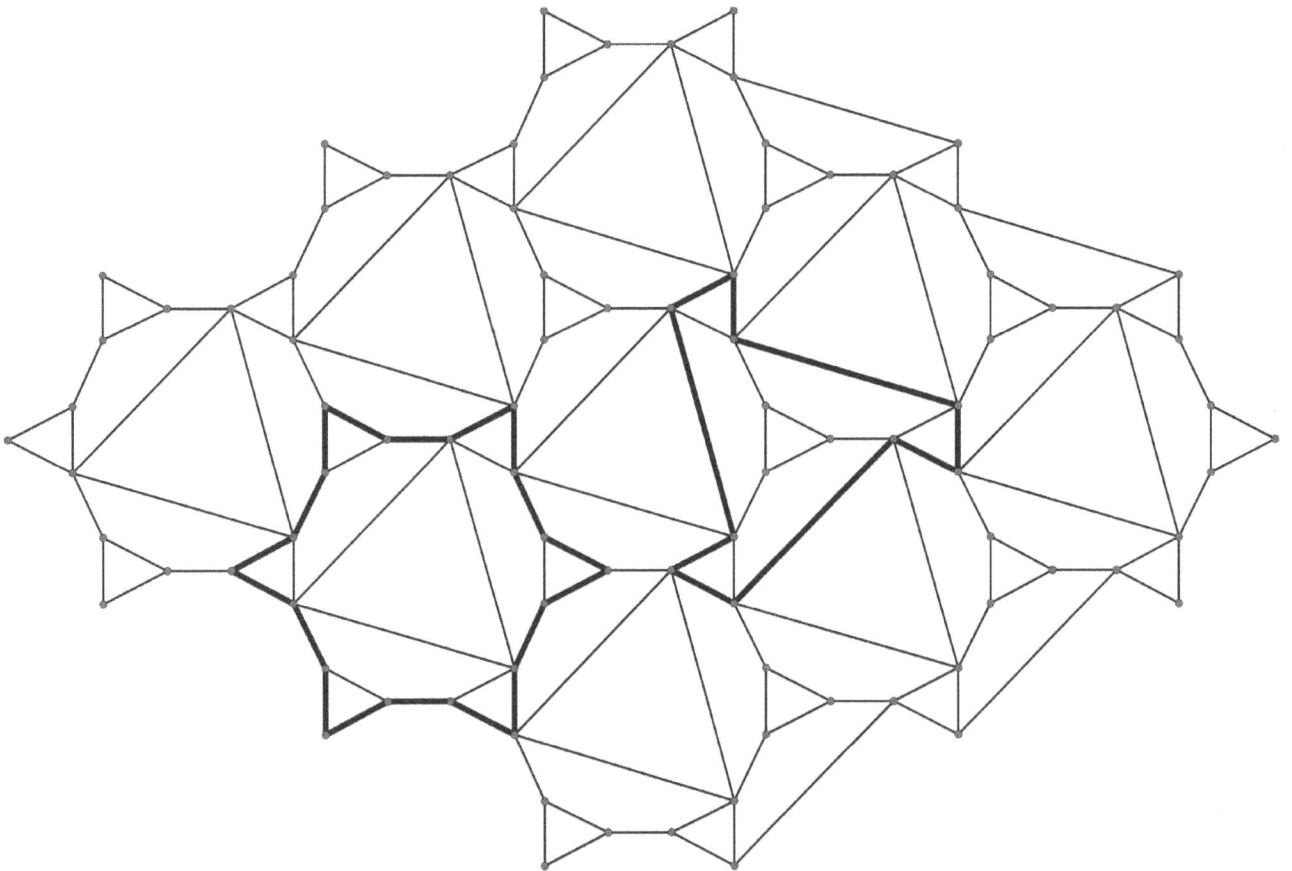

Figure 12.1

The $\prod_{12}$ plane is infinite dimensional and has a free locally nilpotent assignment.

12.2 Scoring configurations

The dodecagon with the inserted large triangle and the extension of this structure with 6 adjacent small triangles, are nilpotent, hence scoring configurations. This latter configuration is illustrated in bold in Figure 12.1. The second configuration indicated in bold in Figure 12.1, we call the *rotor*. The rotor is composite since it has a vertex cover consisting of four small triangles. In fact, the entire board has a small triangle cover and thus is nilpotent. For the purpose of the game, scoring configurations should be restricted to the basic tiles, the dodecagon, the extended dodecagon and the rotor.

12.3 The play

There are 90 points on the playing board so the $\prod_{12}$ game is equipped with 90 numbered discs consisting of 9 groups each consisting of the values {0, ..., 9}. As usual each player draws 10 discs from the boneyard to constitute their hands. One more disc is drawn and placed on a vertex of the large central triangle. Blocking and playing a nonscoring disc in a tile occupied by a dead point are not allowed.

A players turn consists of making a series of scoring plays followed by a nonscoring placement. To improve the pace of the game it is allowed to play two discs in succession in the same turn in order to complete a triangle or pentagon. Players should choose the combination and placement of the two discs in order to obtain further scoring opportunities. A *double play* like this is only allowed to complete a triangle or pentagon; otherwise only single discs can be played at a time.

Due to the double play possibilities, this game has a fast pace and very few dead points occur. Quite often a player can exhaust his/her hand in one turn by making repeated scoring plays.

CHAPTER 13

The Π_{13} game

13.1 The Π_{13} plane

In Figure 13.1 a section of the Π_{13} plane is illustrated.

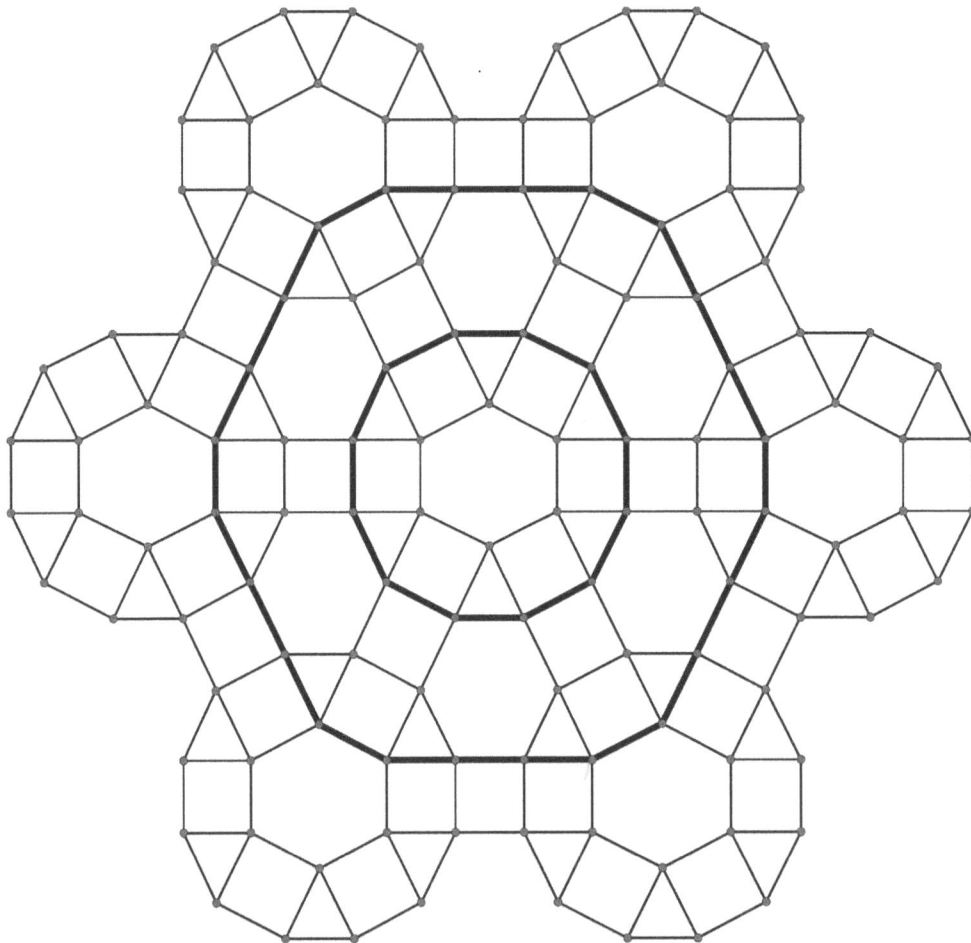

Figure 13.1

The $\prod_{13}$ plane is infinite dimensional and has a free locally nilpotent assignment. It is also *regular of degree 4*. This means that every vertex is incident with exactly four edges. This property allows the definition of two distinct *antiderivatives*. Since each vertex of $\prod_{13}$ has degree 4, there are exactly four regions incident with it. These consist of two pairs which share only the vertex in common. Choosing one of these pairs draw a line segment through the common vertex with an endpoint in the interior of each pair. Starting with the endpoints so constructed repeat this process until every vertex of the plane is covered. The created points and line segments form a graph we call an *antiderivative of $\prod_{13}$* and denote by $\xi(\prod_{13})$. This designation is appropriate in this case since the derivative of $\xi(\prod_{13})$ is $\prod_{13}$. If we use D to denote the derivative operator we have $D\xi(\prod_{13}) = \prod_{13}$.

If the other pair of regions, (which share just a point in common), is selected, then a second antiderivative is obtained. The construction of the two antiderivatives of $\prod_{13}$ is illustrated in Figures 13.2 and 13.3.

Figure 13.2

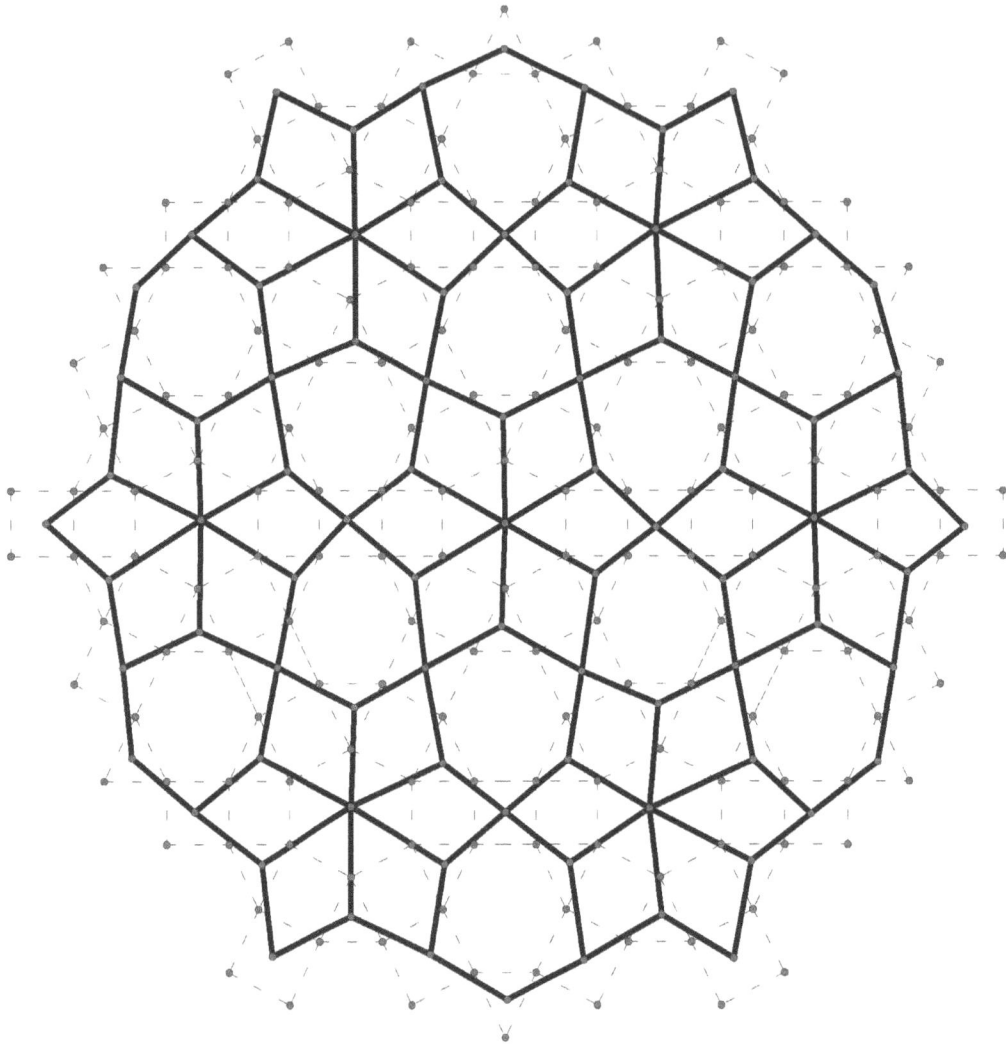

Figure 13.3

Both antiderivatives, indicated in bold in Figures 13.2, 13.3, satisfy the formula $D\xi(\Pi_{13}) = \Pi_{13}$. Note that the construction of the antiderivative depends on the existence of a *2-coloring* in Π_{13}. The regions in Π_{13} can be colored black and white so that no adjacent regions have the same color. The reader interested in graph coloring can consult [4], which also serves as a good introduction to graph theory.

The regions in the same color class form the vertices of the antiderivatives. Note also that the antiderivative in Figure 13.2 is isomorphic to Π_4 and, as we have seen is very rich in nilpotent structures. On the other hand, the antiderivative in Figure 13.3, which we denote by β, has few nonbasic nilpotent polygons. Observe that $\beta = d(\Pi_4)$.

The plane $\prod_4$ also has a 2-coloring so we can also construct its antiderivatives. One of the derivatives is just the triangle lattice which we denote by Δ. In Figure 13.4 the other antiderivative is illustrated. This antiderivative is $\prod_7$ so we have $\xi(\prod_4) = \prod_7$. In this case $D(\prod_7) = \prod_5$ and $D(\Delta) = \prod_1$, so the formula $D\xi(\prod) = \prod$ does *not* hold.

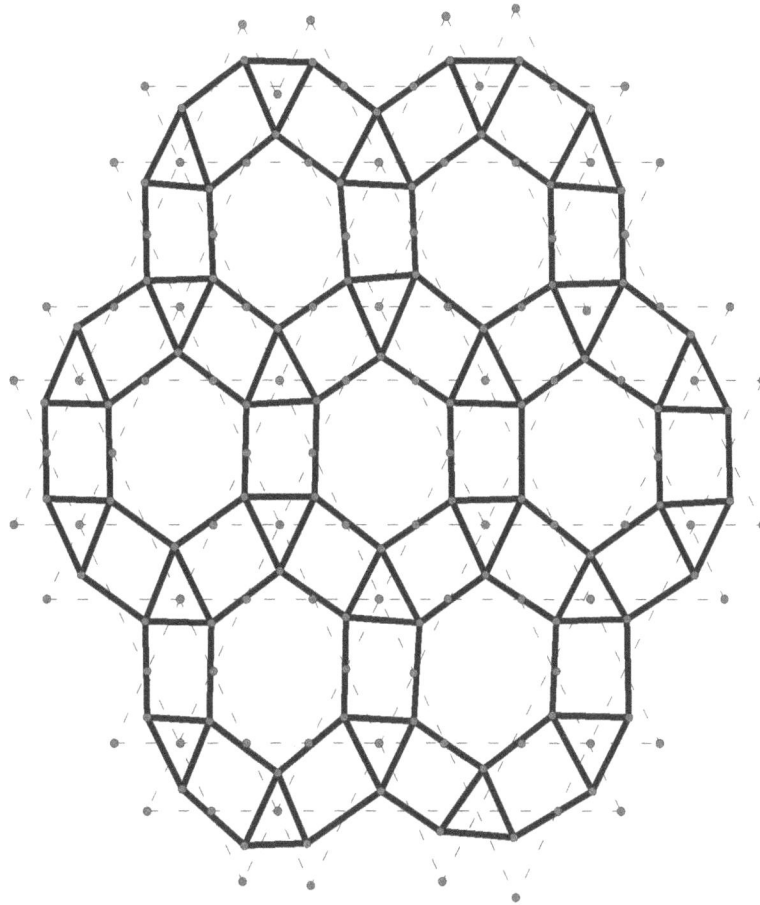

Figure 13.4

13.2 The $\prod_{13}$ game

As in previous games, each player draws 10 discs from the boneyard, and one additional disc is drawn and placed on a point of the central hexagon. The playing board, in Figure 13.1, has 126 points so the boneyard is stocked with 130 points in 13 groups $\{0, \ldots, 9\}$. The two bold dodecagons illustrated in Figure 13.1 are barriers. There should be no blocking and nonscoring discs should not be allowed to be played in a basic tile occupied by a dead point. Scoring configurations should be limited to the ones illustrated in Figure 13.5.

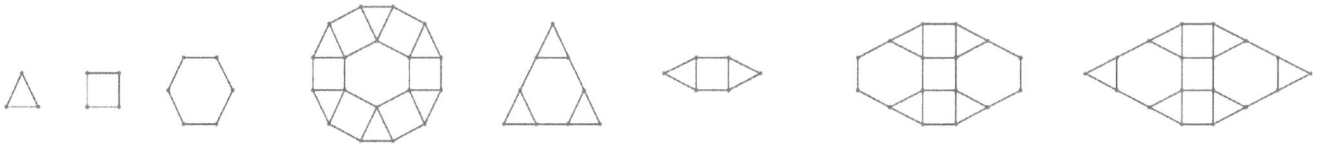

Figure 13.5

The large configuration surrounded by the outer barrier in Figure 13.1 can also be included as a scoring configuration. Otherwise the rules of play are the same as in previous games.

CHAPTER 14

The map of planes

14.1 The map of planes and the $\prod_{14}$ game

A map illustrating the relationship between our planes is illustrated in Figure 14.1.

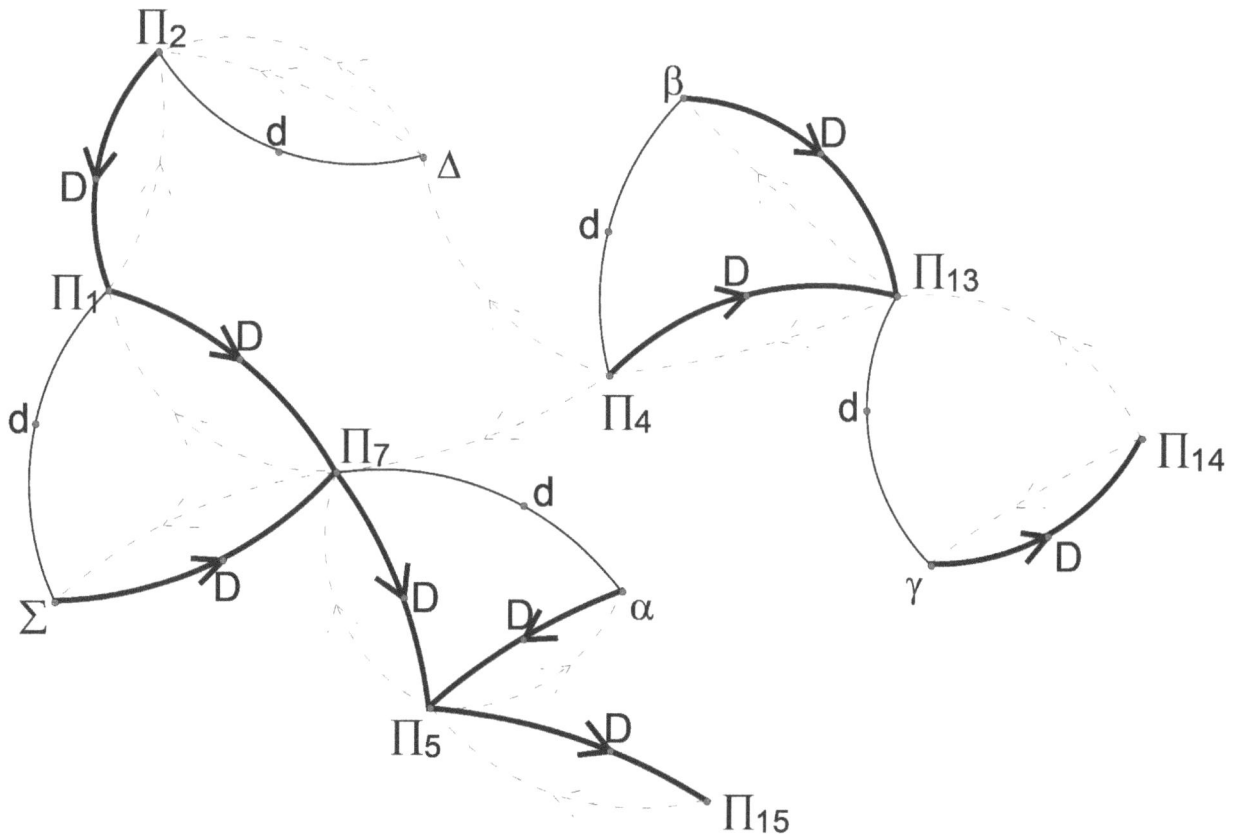

Figure 14.1

In Figure 14.1 the dashed arcs represent the action of the antiderivatives ξ. This Figure is a *directed graph, or digraph.* A digraph G = (V,A) is a collection of points, or vertices, V along with a set A of ordered pairs from V called *arcs*. The ordered pair (a,b) is represented by an arc directed from a to b. In Figure 14.1 the arcs labeled d are not provided with arrows. Since $dd(\Pi) = \Pi$, there should be two d arcs, one in either direction. We have abbreviated this representation by using just a single undirected edge.

Some observations regarding this Figure are the following. Assume that Π is a plane tiled edge to edge by convex polygons.

(1) $\xi D(\Pi) = \Pi$ holds for every plane Π.
(2) $D(\Pi)$ is regular of degree 4 for every plane Π. Thus if Π is not 4-regular then it cannot be the derivative of any plane, (Π_4 is an example.)
(3) In case Π is regular of degree 4, then $D\xi(\Pi) = \Pi$.
(4) $D(\Pi)$ always has two antiderivatives.
(5) $Dd(\Pi) = D(\Pi)$. (Note that a D arc from Δ to Π_1 is missing in Figure 14.1.)
(6) Π has antiderivatives iff the vertices of Π have degree 4 or 6.
(7) If Π is infinite dimensional then so is $D(\Pi)$, and if Π has a free locally nilpotent assignment then so does $D(\Pi)$.

The existence of the antiderivatives in (6) is dependent upon the existence of a 2-coloring of the regions of Π. In the following Theorem we show that such a 2-coloring always exists for a plane graph G provided every vertex of G has even degree. The *chromatic number $\chi(G)$ of a graph* is the minimum number of colors which can be used to create a proper coloring of the vertex set of G. A graph G is *bipartite* in case $\chi(G) = 2$.

Theorem 14.1 *Let Π be a plane tiled edge to edge by polygons. Suppose also that every vertex of Π has even degree larger than or equal to 4. Then the regions of Π are 2-colorable. This is the same as saying that $d(\Pi)$ is bipartite.*

Proof. Let G denote a maximal 2-colorable subgraph of Π and suppose G does not cover all the regions of Π. Then there exists a tile T of Π which lies in the exterior of G but contains a boundary edge of G. If all the edges of T which lie in G bound regions of G all with the same color, then T can be colored with the opposite color. But this would contradict the maximality of G.

Suppose then that the edges of T in G correspond to regions of G which have both colors. Then two such edges e_1, e_2 can be found which are adjacent on the boundary of G. If x is the common point between these two edges then since the degree of x is even, there are an odd number of colored regions incident with x. But in that case e_1, e_2 correspond to regions of G with the same color, contradicting our choice of these edges.

The conclusion is that G covers the entire plane $\prod$ and thus $\prod$ is 2-colorable. $\square$

Note that all the game planes in Figure 14.1 are generated from $\prod_4$ via the operators D, d, ξ. This digraph and all extensions form an infinite digraph which we denote by $\langle \prod_4 \rangle$ since it is generated by $\prod_4$. The plane $\prod_6$ does not lie in $\langle \prod_4 \rangle$, although it is tiled in the same way as the planes in $\langle \prod_4 \rangle$ with hexagons triangles and quadrilaterals. If we let θ denote the digraph which includes all planes tiled only by convex hexagons triangles and quadrilaterals and whose arcs are determined by the actions of D, d and ξ, then $\langle \prod_4 \rangle$ and $\langle \prod_6 \rangle$ would constitute two connected components of θ. This observation leads to the following interesting question.

Question 14.2 *How many connected components does the digraph θ have?*

Note that a *component* of a graph/digraph G is a maximal connected subgraph/subdigraph of G.

14.2 The $\prod_{14}$ game

The section of the $\prod_{14}$ plane to be used as the board for the $\prod_{14}$ game is illustrated in Figure 14.2. The $\prod_{14}$ is infinite dimensional and has a free locally nilpotent assignment. The scoring configurations suggested for the $\prod_{14}$ game are illustrated in Figure 14.3. As usual these configurations correspond to nilpotent polygons in the $\prod_{14}$ plane.

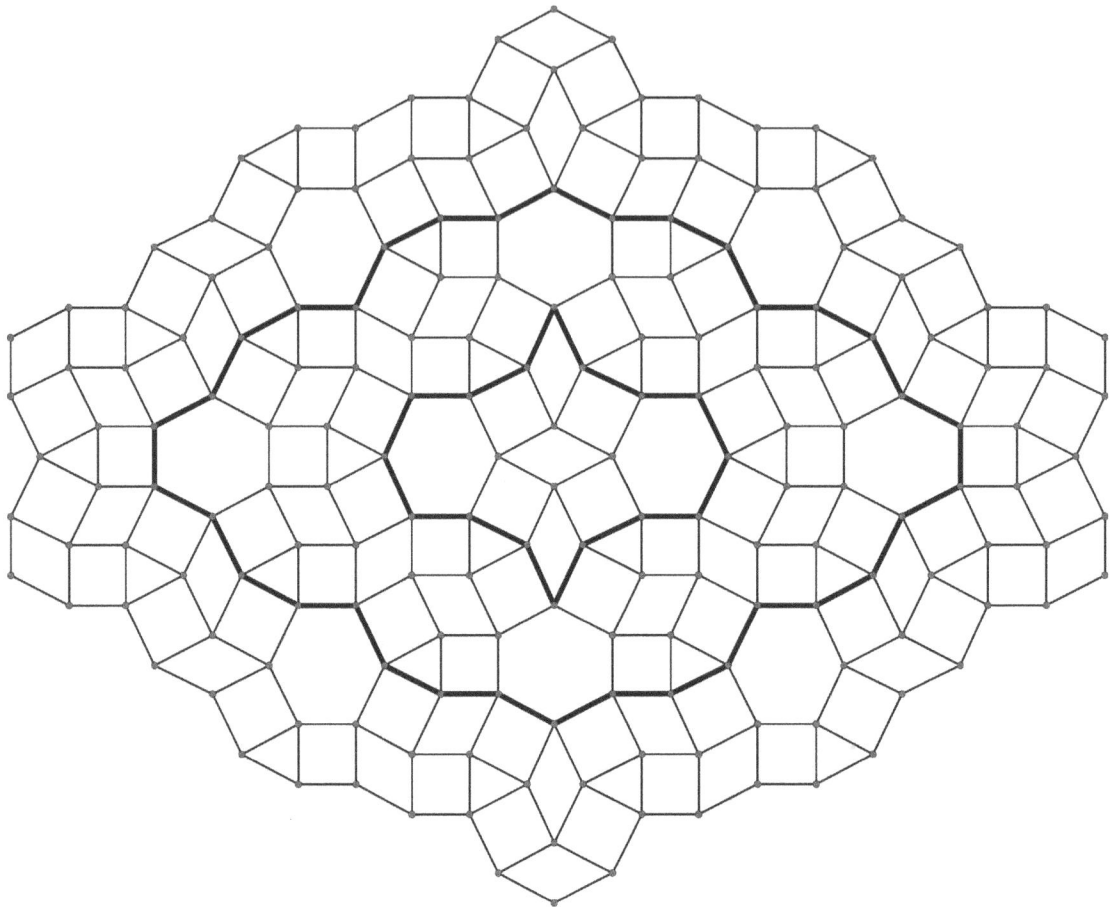

Figure 14.2

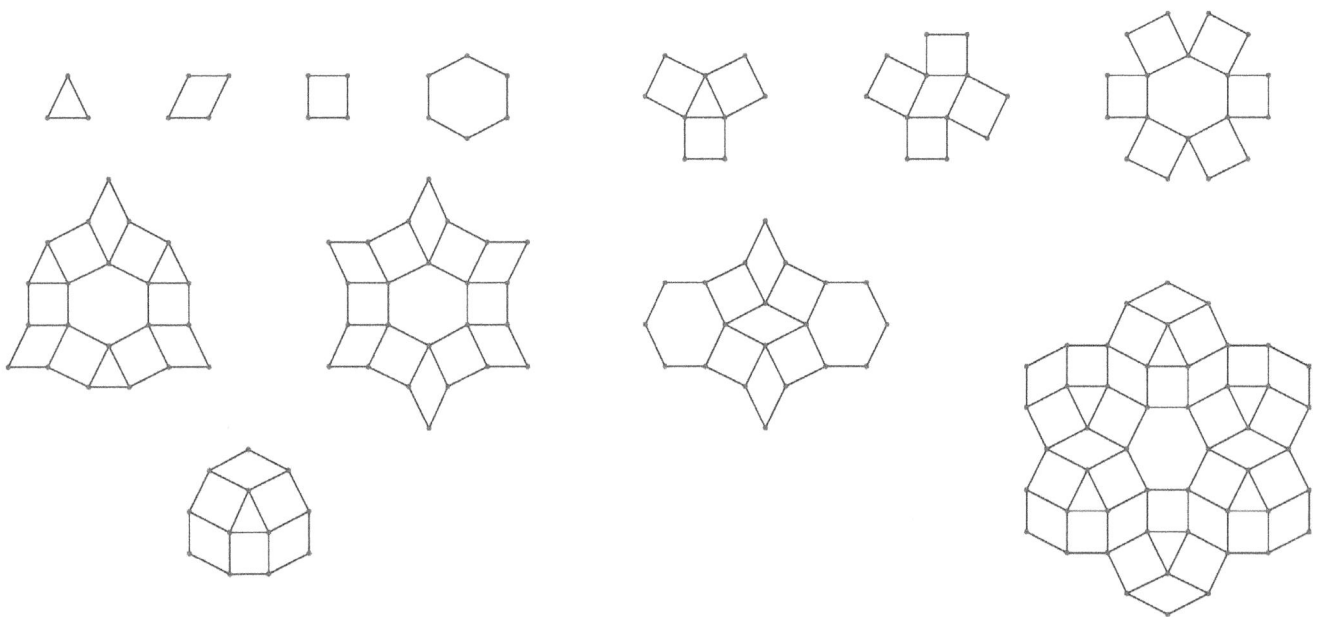

Figure 14.3

The game is played with the two barriers indicated in bold in Figure 14.1. There should be no blocking, and playing a nonscoring disc in a basic region occupied by a dead point should be disallowed. There are 180 points in the playing board so the game is accompanied by 18 groups of numbered discs, each group consisting of the values {0, …, 9}. Each player draws 10 discs from the boneyard, and then one more disc is drawn and placed on one of the points of the central rhomb. The rules of play are otherwise the same as in previous games.

CHAPTER 15

The Π_{15} game

15.1 The Π_{15} plane

In Figure 15.1 a section of the Π_{15} plane is illustrated in bold. This plane is obtained as the derivative of Π_5. (See Figure 14.1.) It is infinite dimensional and has a free locally nilpotent assignment.

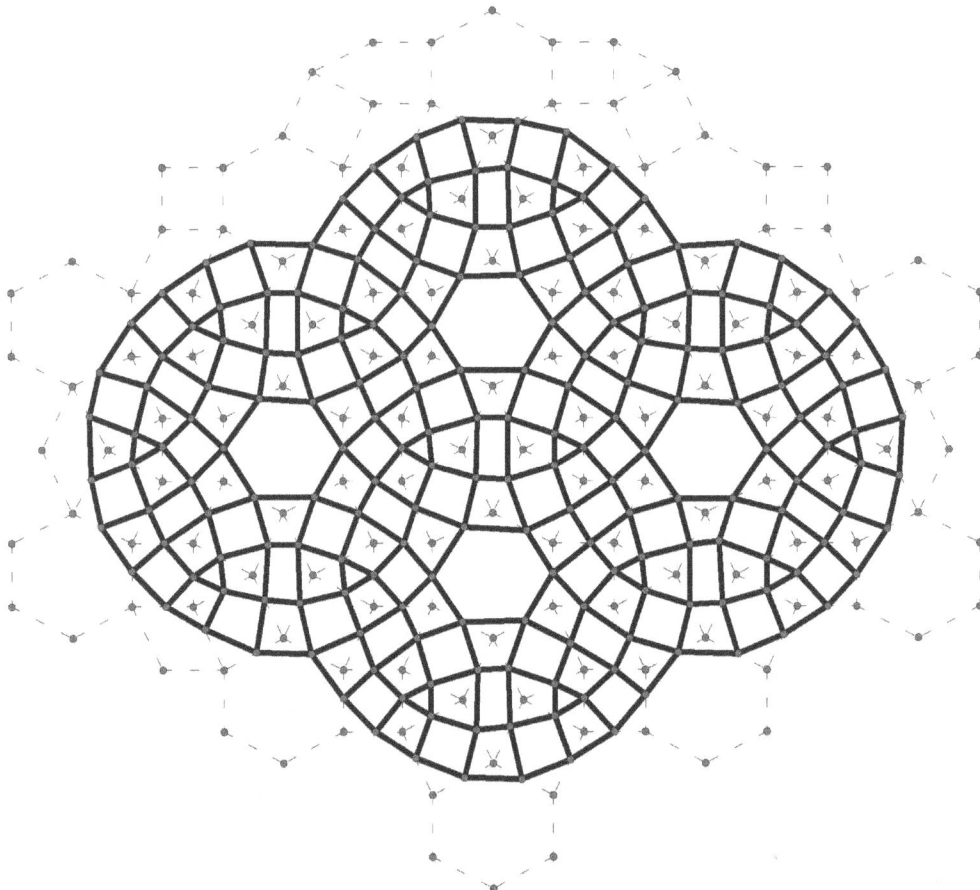

Figure 15.1

15.2 Nilpotent $\prod_{15}$ polygons

In Figure 15.2 some recommended scoring configurations for the $\prod_{15}$ game are illustrated. As usual these correspond to nilpotent $\prod_{15}$ polygons.

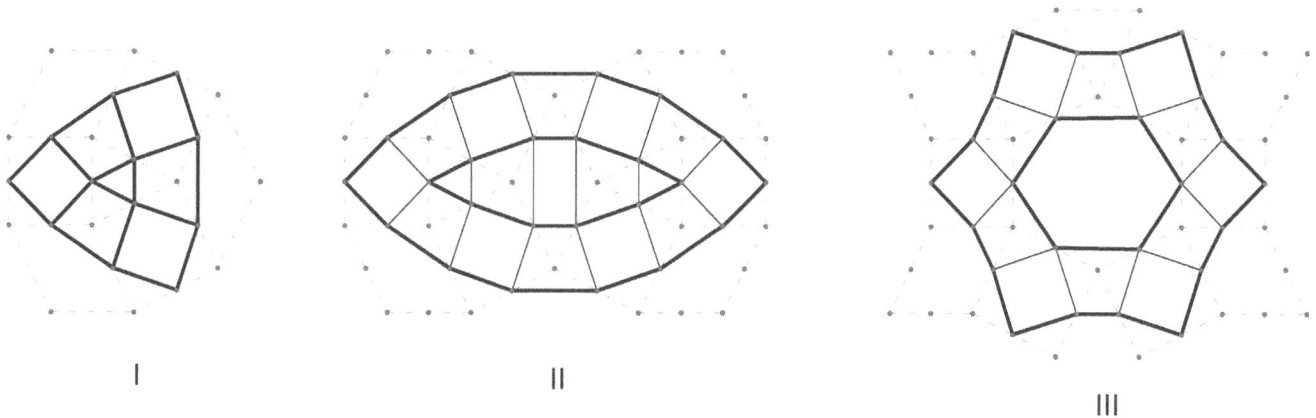

I II III

Figure 15.2

In Figure 15.2(II) two scoring configurations are illustrated: the *small oval* and the *large oval*. The large oval which occurs at the center of the playing board in Figure 15.1 should be a barrier. The two *rings* in Figure 15.3(I) and the *tripods* in Figure 15.3(II) should also be included as scoring configurations.

A second barrier should be constructed which contains the four hexagons and arcs between them. Blocking and playing nonscoring discs in a region occupied by a dead point should be disallowed.

15.3 The play

The playing board has 192 points so the game is accompanied by 190 discs in 19 groups each with values $\{0, \ldots, 9\}$. Each player draws 10 discs from the boneyard and another disc is drawn and placed on one of the points of the central rectangle. Play then proceeds as in previous games.

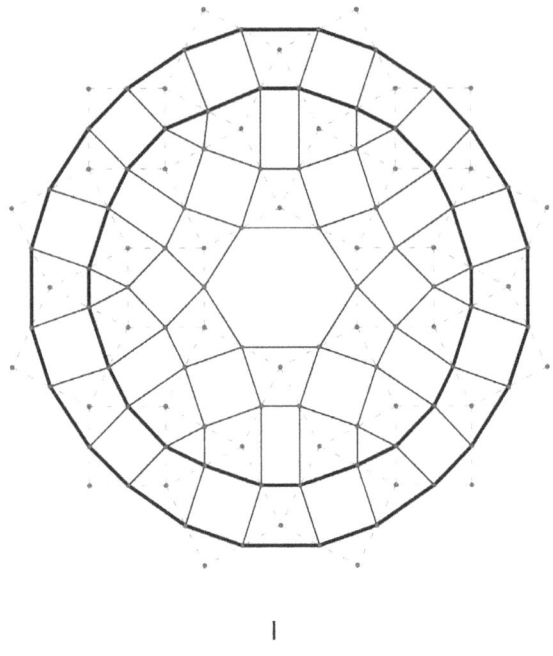

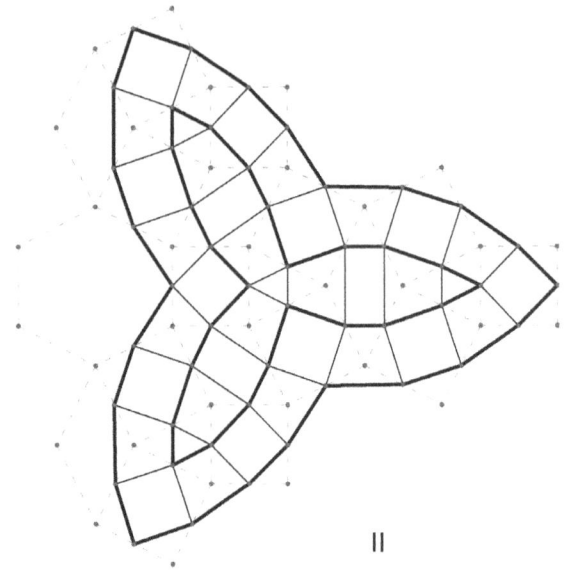

I

II

Figure 15.3

CHAPTER 16

The Π_{16} game

16.1 The Π_{16} plane

A section of the Π_{16} plane is illustrated in Figure 16.1. It is the derivative of Π_{10}. The basic tiles consist of triangles, various quadrilaterals and pentagons.

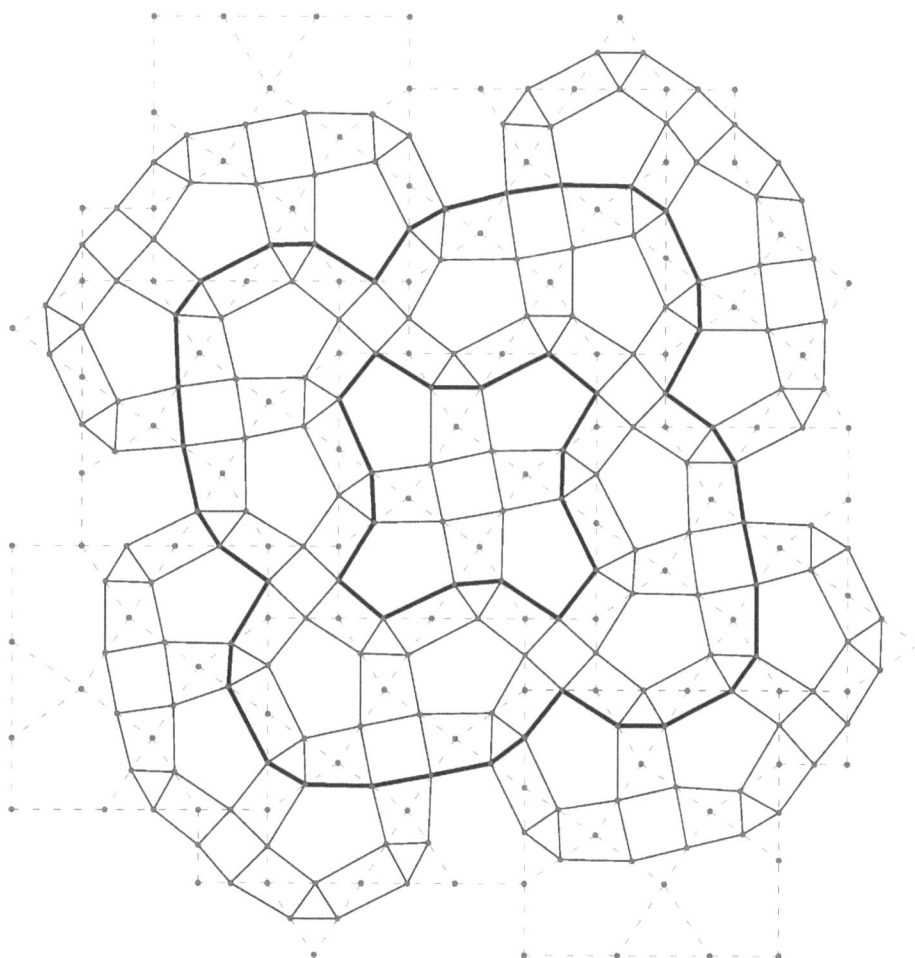

Figure 16.1

16.2 Nilpotent $\prod_{16}$ polygons

Besides the basic tiles, scoring configurations for the $\prod_{16}$ game are illustrated in Figure 16.2.

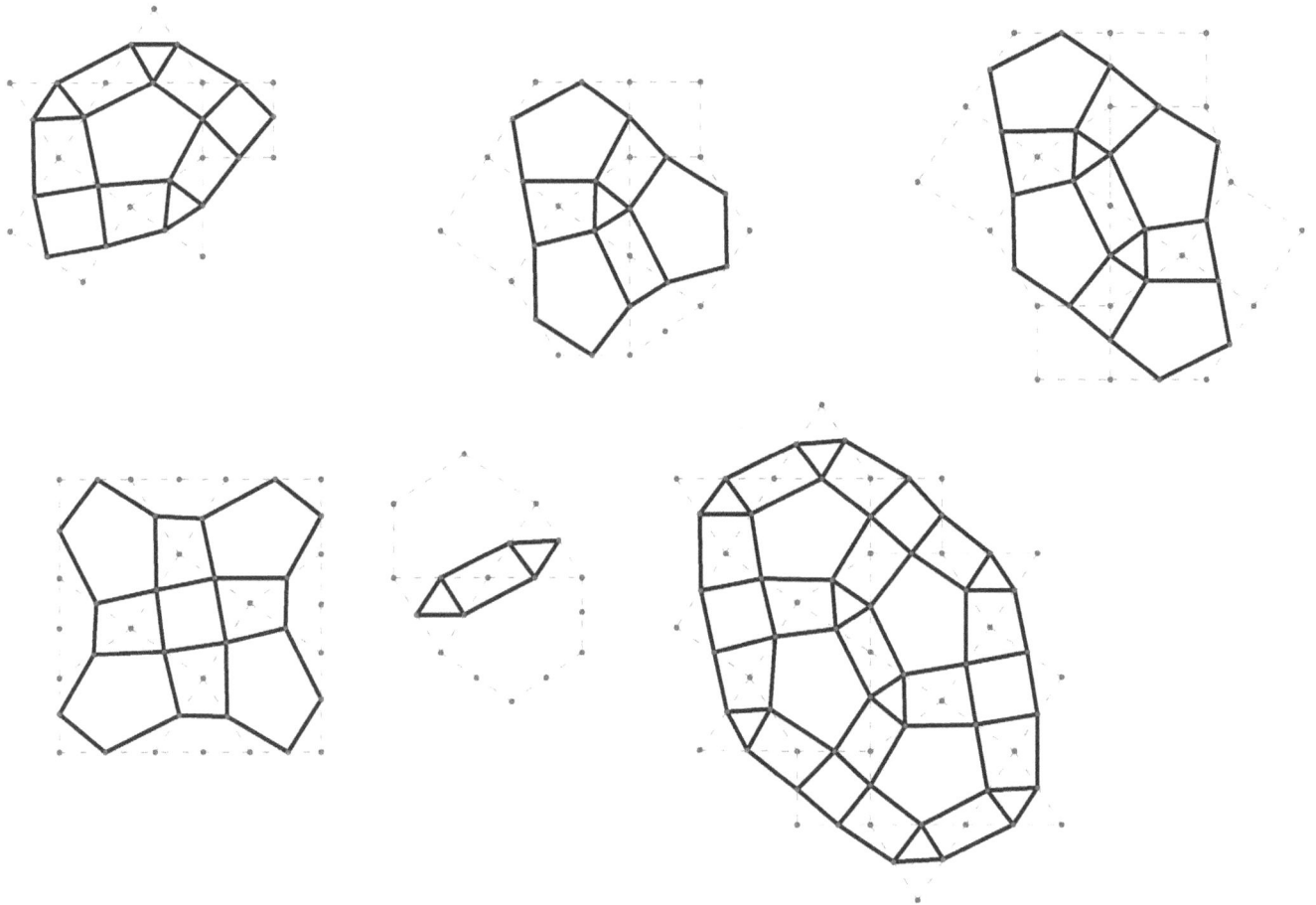

Figure 16.2

16.3 the play

The game board has 176 points and is equipped with 180 numbered discs in 18 groups each with values {0, ..., 9}. Blocking and playing a nonscoring disc in a basic region occupied by a dead point should be disallowed. Otherwise the play of the game proceeds as in previous games.

CHAPTER 17

The $\prod_{17}$ game

17.1 The $\prod_{17}$ plane

The $\prod_{17}$ plane is the derivative of the plane illustrated in Figure 17.1.

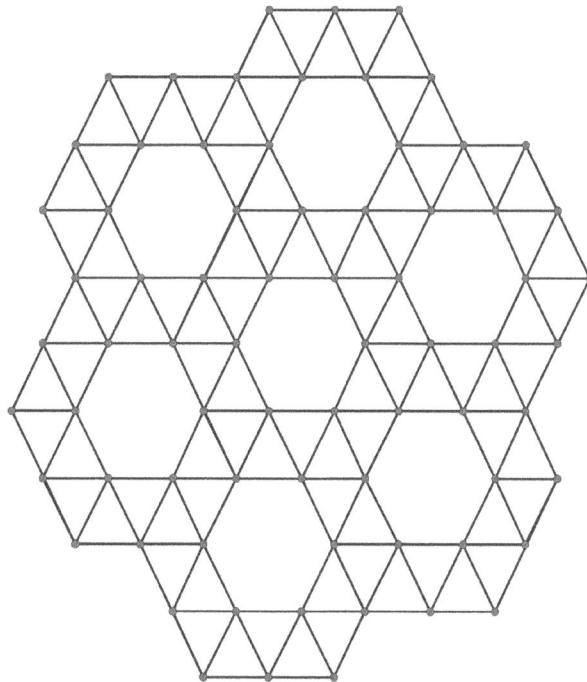

Figure 17.1

We have avoided using planes with adjacent triangles since opposite points of the rhomb so formed must have the same value. This usually results in a plane with no free locally nilpotent assignment. This is true of the plane illustrated in Figure 17.1. The derivative of such a plane, however, can yield very interesting planes. The derivative $\prod_{17}$ of the plane in Figure 17.1 is illustrated in Figure 17.2.

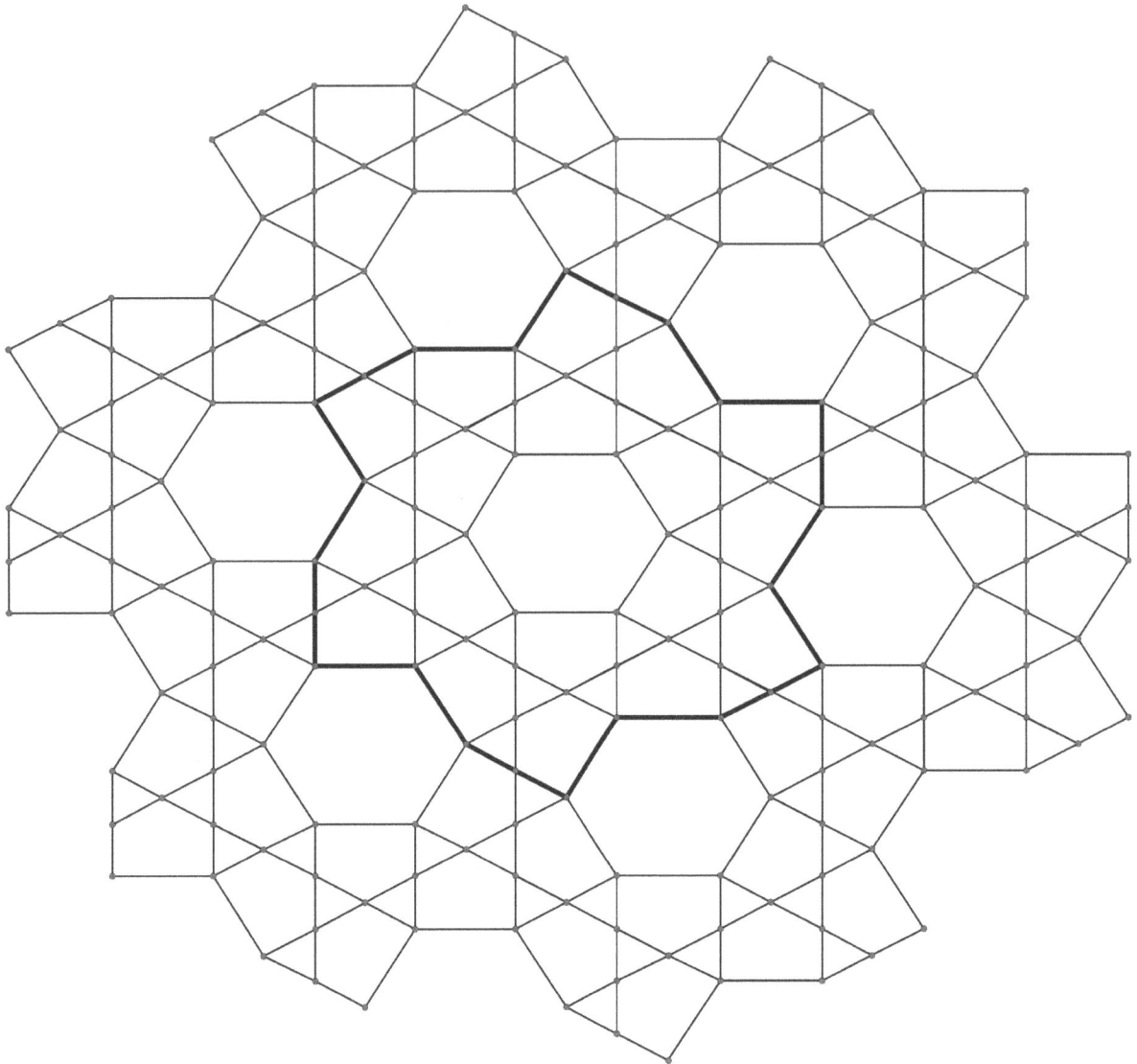

Figure 17.2

The $\prod_{17}$ plane is infinite dimensional and has a free locally nilpotent assignment.

17.2 Nilpotent $\prod_{17}$ polygons

Besides the basic tiles: the hexagon, the triangle and the pentagon, the two hexagons in Figure 17.3 are scoring configurations.

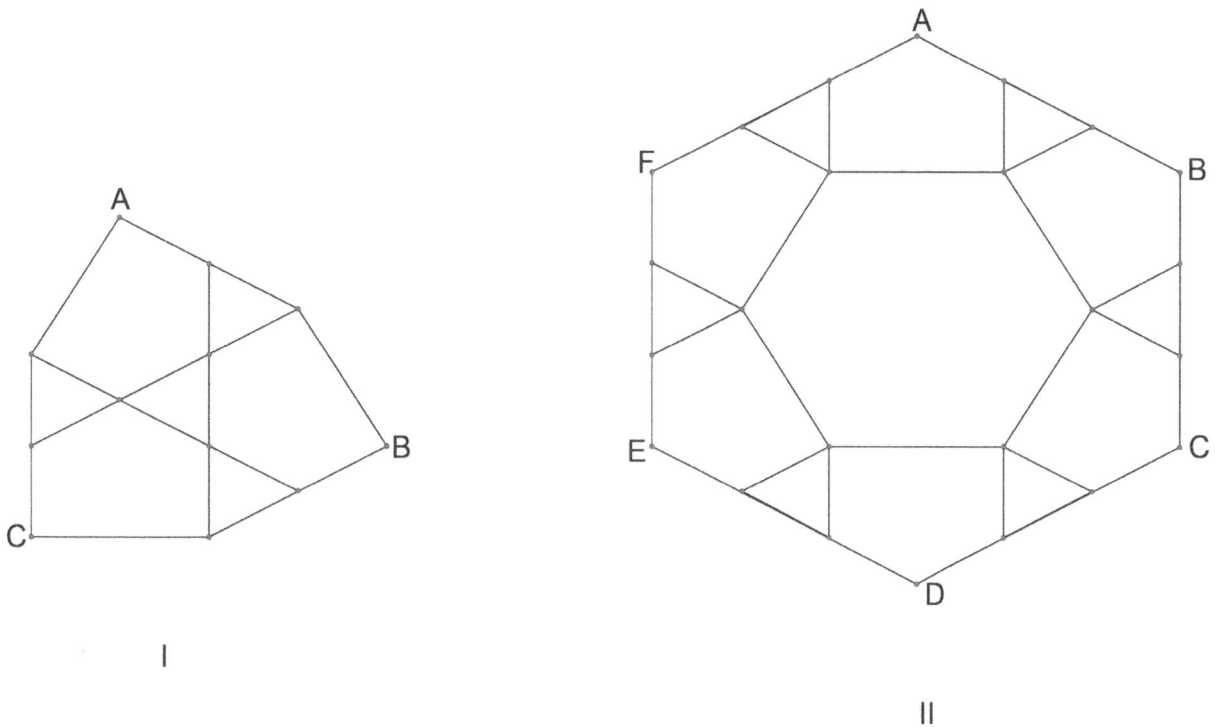

Figure 17.3

The hexagon in Figure 17.3(I) will be denoted by N. It is nilpotent since it is a side extension of the triangle. The hexagon H_3 in Figure 17.3(II) is nilpotent since it is a side extension of the basic hexagon. The following two observations we leave for the reader to prove.

Theorem 17.1 *In any locally nilpotent assignment the corners A, B, C of N sum to 0, and the corners A, B, C, D, E, F of H_3 also sum to 0.*

In Figure 17.4 some more scoring configurations are illustrated. The nilpotence of these figures is a consequence of the following.

Theorem 17.2 *In Figure 17.4(I) four triangles have been appended to H_3. They involve the addition of four points. In any locally nilpotent assignment the sum of the values assigned to these four points is 0. The nilpotence of the polygon in Figure 17.4(II) follows from symmetry with or without the dashed triangles.*

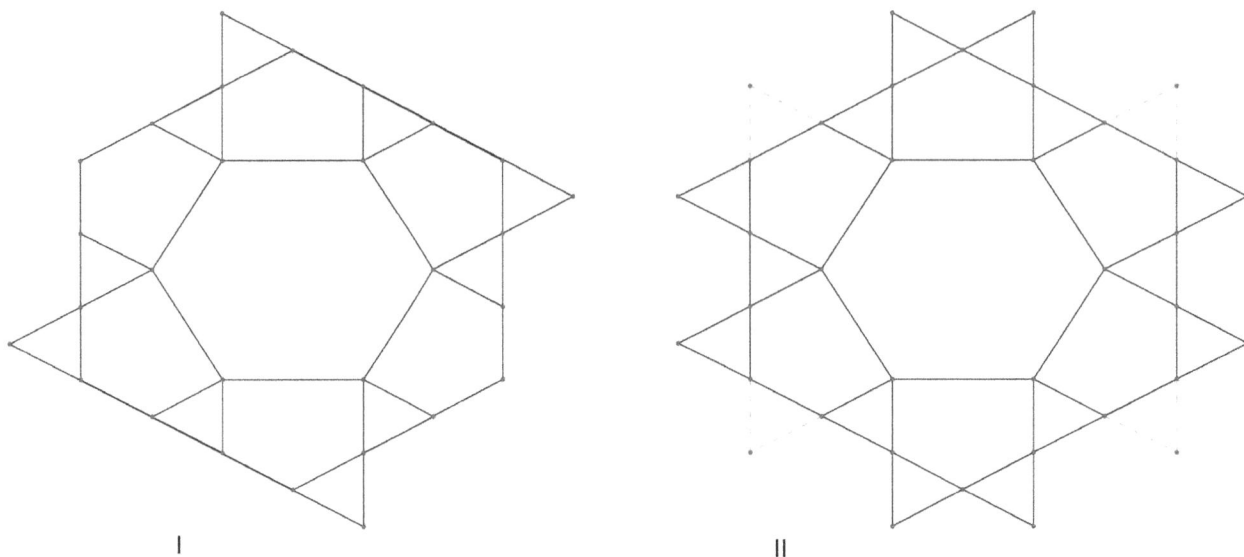

Figure 17.4

In Figure 17.5 a few more scoring configurations are illustrated.

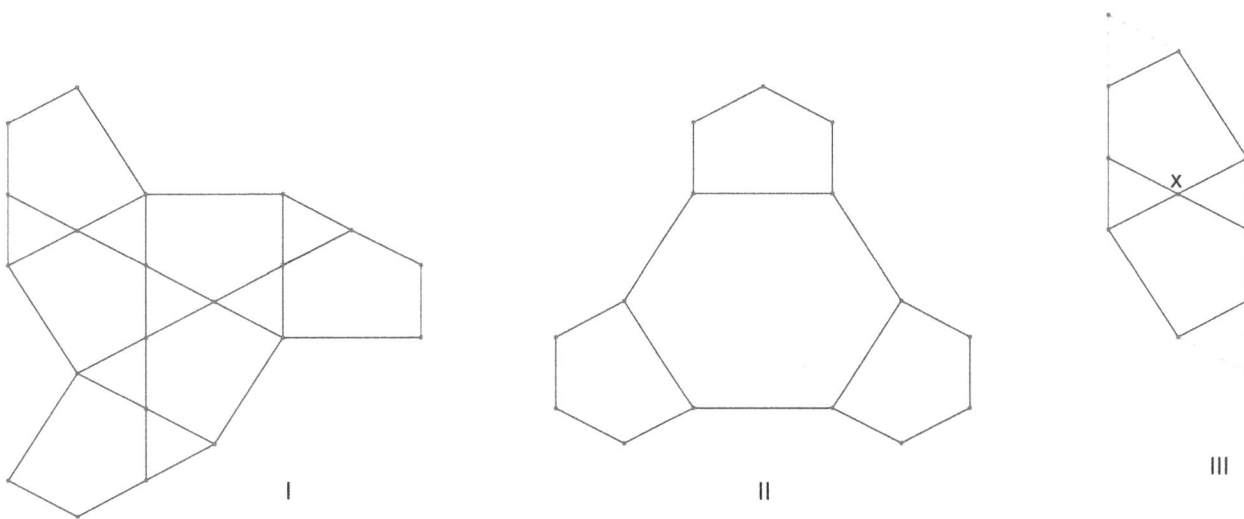

Figure 17.5

The polygon in Figure 17.5(II) is a side extension of the basic hexagon. In Figure 17.5(III) the illustrated hexagon, with or without the dashed edges, is conditionally nilpotent with x = 0 or x = 5.

Thus far all the illustrated $\prod_{17}$ scoring configurations are prime. In Figure 17.6 two composite scoring configurations are illustrated.

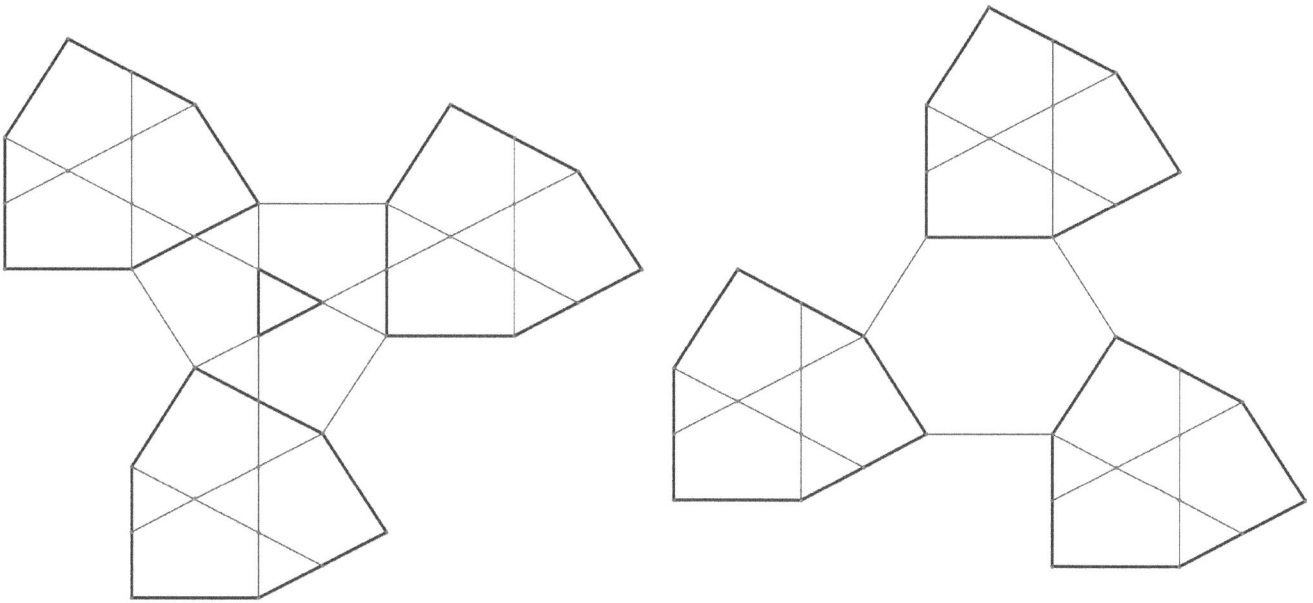

Figure 17.6

The region indicated in bold in Figure 17.2, which we call a *galaxy,* is also a nilpotent $\prod_{17}$ polygon. This fact we leave for the reader to demonstrate. Due to its size, it occurs infrequently in the play of the $\prod_{17}$ game. In Figure 17.7 a nilpotent subgraph of the galaxy is illustrated. In this graph the indicated points A, B, C, D sum to 0(mod 5) in any locally nilpotent assignment.

The entire gameboard as illustrated in Figure 17.2, is a *cluster* of 7 galaxies we denote by G^7. To prove the nilpotence of G^7 we proceed in the usual way of covering as much of the graph as possible with small nilpotent polygons as in Figure 17.8. The nilpotence of G^7 will be proved once it is shown that the values given to the points A, B, …, R in any locally nilpotent assignment sum to 0(mod 5). One method for accomplishing this task involves a successive replacement of these points by points with the same sum (mod 5). Several of these replacements are illustrated in Figure 17.9.

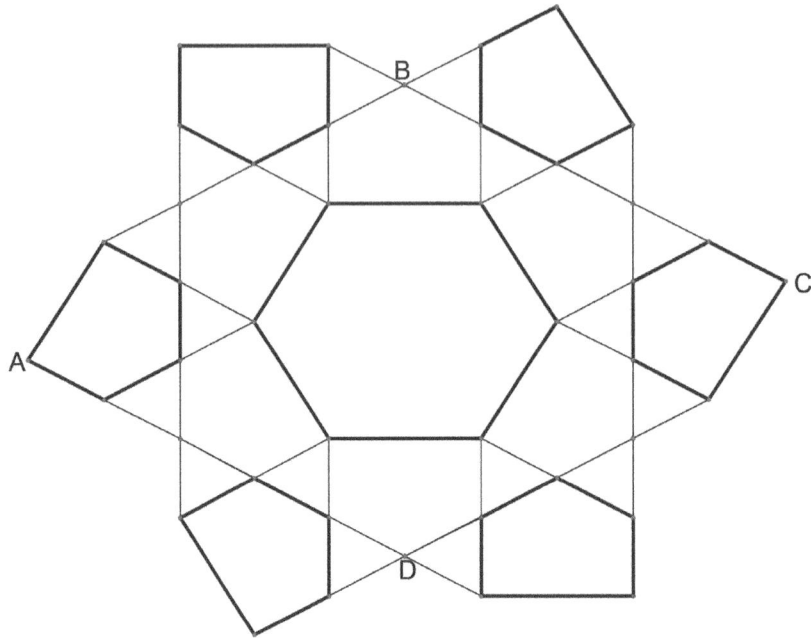

Figure 17.7

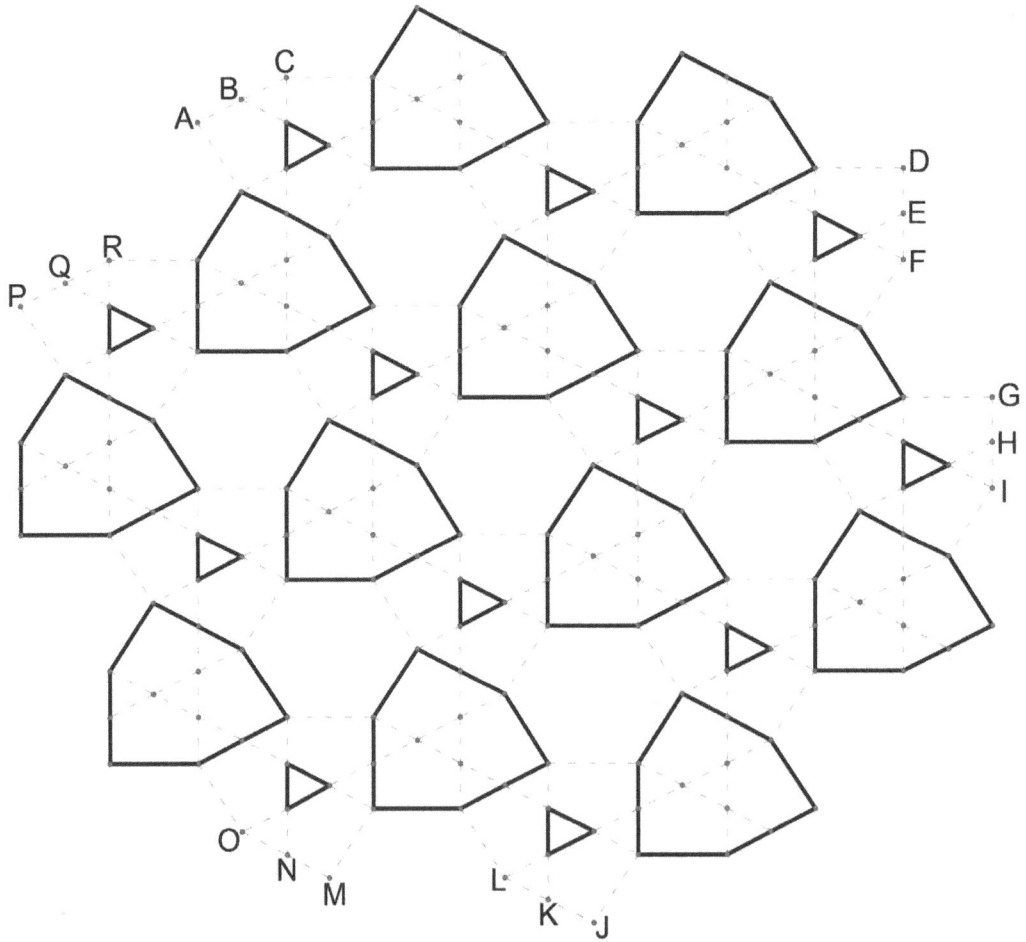

Figure 17.8

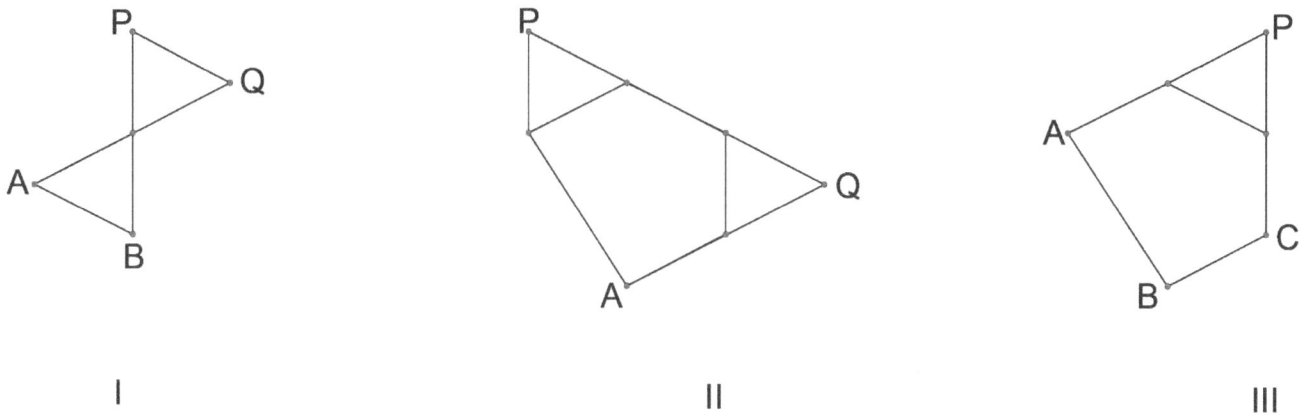

Figure 17.9

Whenever a replacement results in a set of points with 0-sum, the set of points can be removed. When all points are removed the 0-sum for the original set of points is proved. This *replacement method* can be used to show that the galaxy is nilpotent.

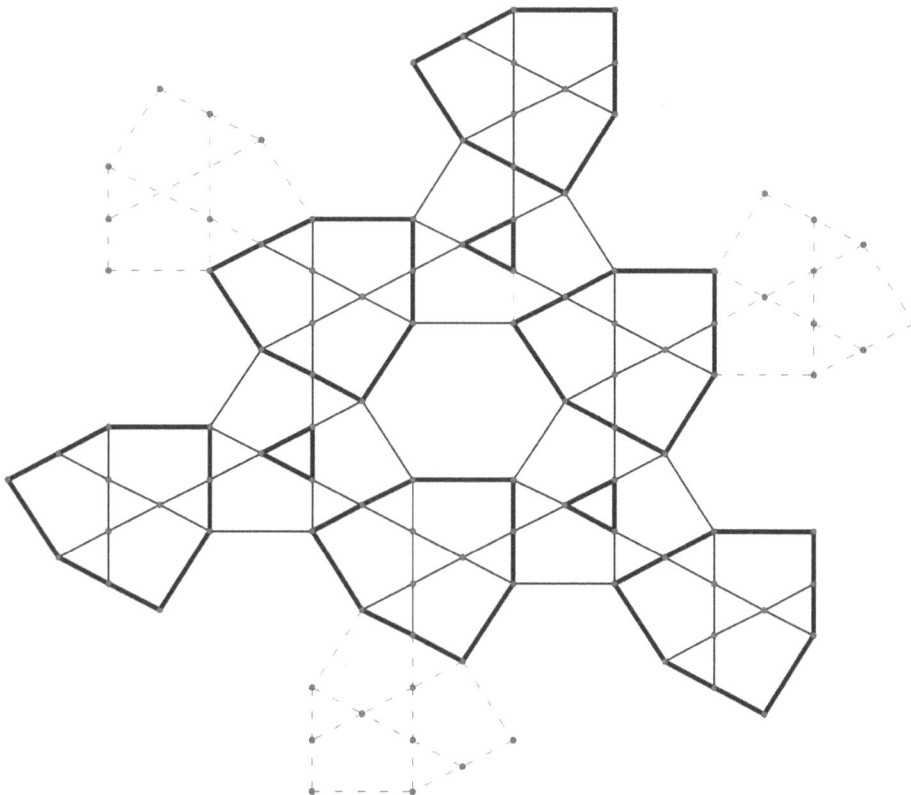

Figure 17.10

In Figure 17.10 the undashed portion is immediately seen to be nilpotent by virtue if the indicated dissection in bold. From the nilpotence of the central galaxy we can conclude that the added points belonging to the appended bold N-polygons have 0-sum. Then from symmetry we can conclude that the entire polygon in Figure 17.10, with dashed edges included, is nilpotent.

Now count all the points in each of the six galaxies in G^7 and let Ω denote the polygon in Figure 17.10. Since each galaxy is nilpotent, the sum is 0 in any locally nilpotent assignment. But in this count all the points in Ω are counted more than once. To be precise, the points in Figure 17.11 are counted three times since they belong to three galaxies, and the remaining points in Ω are counted exactly twice.

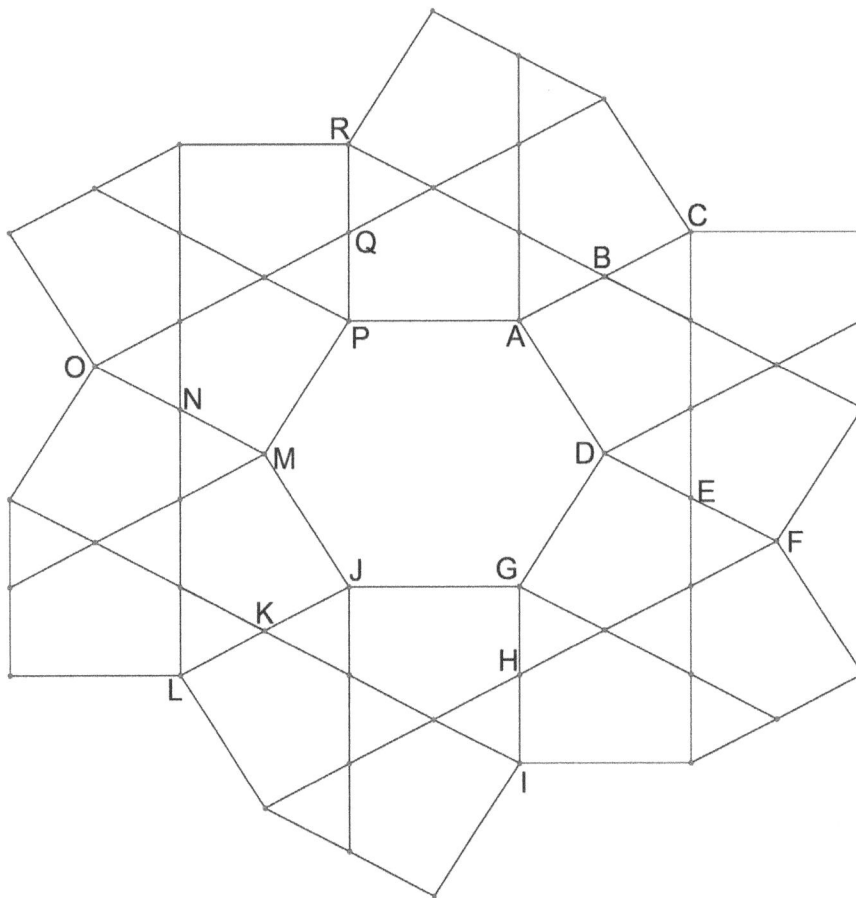

Figure 17.11

The values given to the labeled points in Figure 17.11 are easily seen to have sum 0 in any locally nilpotent assignment. Now since Ω is also nilpotent, it follows that the entire G^7 galaxy cluster is nilpotent. The labeled points in Figure 17.8 must then have sum 0, so it should be possible to show this by the method of replacement. It is however a difficult task.

17.3 Play of the $\prod_{17}$ game

The boundary of the central galaxy of the gameboard, (in Figure 17.2), should be used as a barrier. The basic scoring configurations consist of the triangle, hexagon and pentagon. The polygons N, H_3 in Figure 17.3 and the polygons illustrated in Figure 17.5 should be included as scoring configurations. The extensions of H_3 in Figure 17.4 are a little hard to recognize and also take away from the value of achieving an H_3, so we recommend that these not be considered as scoring configurations. The composite polygons in Figure 17.6 occur frequently and should be counted as scoring configurations.

Blocking should be prohibited and playing a nonscoring disc on the boundary of a region occupied by a dead point should also be disallowed.

The playing board has nearly 200 points, so the game is accompanied by 200 discs numbered {0, ... , 9) in 20 groups. Each player draws 10 discs from the boneyard, and one extra disc is drawn and placed on a point of the central hexagon. Play then continues as usual.

REFERENCES

[1] R. Fletcher, Tilings with Hexagons and Halfstars, Westwood Books Publishing, 2022

[2] R. Fletcher, Game of Rhombs / Game of Triangles, Westwood Books Publishing, 2022

[3] B. Grunbaum and G.C. Shephard, Tilings & Patterns, 2nd ed., Dover Publications, Inc., N.Y., 2016

[4] R. Wilson, Graphs, Colourings and the Four-color Theorem, Oxford University Press, 2002

INDEX

A

Alternating Hexagon 36

B

basic tiles vi
blocking play 61, 62, 125, 128, 133
boat 135
boneyard vii
box 142

C

chevron 3
circuit 5
composite polygon 187
conditional nilpotence 15, 144
conflict 43
congruent to 0 mod 5 vii
contextual nilpotence 44

D

dead point 20, 124
derivative 153
diagonal of a polygon 5
diamond curtain 94
dimension 49, 138
dodecagon 134
double play 151, 163
double score 21, 61, 124, 129

E

edge v
edge couple 29
edge set v

equivalent region 49
even polygon 5
extensions 11

F

factorization 9, 82
free locally nilpotent assignment 51, 132
fundamental region 49

G

graph v

H

H_3 pattern 84
happyman 156
hexagon H(a,b,c,d,e,f) 10

I

inlet 150
intrinsic nilpotence 16
isomorphism 154, 159
isosceles hexagon 120, 122
isosceles trapezoid T(a,b,a+b) 29

L

linear pattern 27
locally nilpotent assignment v

M

mites 25
monohedral tiling vi

N

near regular hexagon 33, 122
nilpotence v
nilpotent quadruple 44
nonscoring disc 125
nonscoring position 18
nonsegment row 77

P

parallelogram 96, 99
parallelogram conglomerate 105
parallelogram reduction 46
path 5
pattern 31
pentagon 75, 104
periodic tiling 2
planar graph v
plane v
point couple 26
point extension 25, 53
polyhex 64
prime polygon 31
propeller 156
proper tiling 149

R

region v
rhomb 49

S

scoring configuration vi
scoring position 18, 129
second derivative 153
segment 5
semiregular polygon 115
side extension 155
simple 0-value dissection 89
size of a circuit 10
soft polygon 15
space tiling 142
stream 127
subgraph 2

T

tetradecagon 132
tiles v
tiling v
trapezoid 27, 47
trapezoid reduction 44
trapezoids 82
triangles 26
tripod 25, 54
tripod and M6 pattern 31

V

vertex set v

W

wedge 72
wedge and rhomb pattern 49
wedge complex 123

www.ingramcontent.com/pod-product-compliance
Lightning Source LLC
Chambersburg PA
CBHW052342210326
41597CB00037B/6220